Final Years on the Tractor Seat

Also by Arthur Battelle

Early Years on the Tractor Seat
More Years on the Tractor Seat

Final Years on the Tractor Seat

Arthur Battelle

Farming Press

First published 1996

ISBN 0 85236 355 9

A catalogue record for this book is available from the British Library

Published by Farming Press
Miller Freeman Professional Ltd
Wharfedale Road, Ipswich IP1 4LG, United Kingdom

Distributed in North America
by Diamond Farm Enterprises,
Box 537, Alexandria Bay, NY 13607, USA

Text cartoons by Jim Eckersley
Cover design by Mark Beesley
Typeset by Galleon Typesetting, Ipswich
Printed and bound in Great Britain by Biddles Ltd,
Guildford and King's Lynn

Preface

Final Years on the Tractor Seat sounds as if I am already dead, but in this year of 1996, even though the magical three score and ten years have been reached, I find myself very much alive, with plenty of ambition left and the physical ability to do just a little more.

Many of you will recall the two previous books in my tractor trilogy, each one a complete story in itself, as indeed this one is. Looking back through the story of a life with tractors, I can see in *Early Years on the Tractor Seat* younger years a light-hearted approach to reality, even during the war.

As life progressed and became more serious, age changed my outlook, as shown in *More Years*, and in this third volume, I am in at the deep end, struggling to run two businesses and act as the English-speaking representative for the world for another company. The years flash past and the pace of life changes but the smiles in the story are still there as earlier. And as I read it over, I wonder how ever a country lad managed to do these things!

Arthur Battelle

This book is dedicated to the small band of people who in the middle years of this century left their loved ones, often for long periods, to serve the interests of agriculture and the British tractor industry in particular, returning with hope for the future and a few coins to jangle in their pockets.

1

AS 1960 drew to a close and 1961 took over, I could look back on a year of working for Ford Motor Company – travelling through Sweden for six months with a tractor demonstration team, introducing the Fordson Super Major to all the Fordson dealers in Europe at a very intensive and expertly staged event in Hamburg, coming home to the UK with our demonstration trucks and working in an office for the first time in my life. What the Fordson office staff in Essex thought of this uneducated yokel in their midst I do not know, but they were helpful to me and never derogatory . . . Ford democracy I suppose. (At this time Ford was a British company, Ford in America having decided or indeed been forced, during the war, to relinquish control of most European companies previously owned by them. This would change later – but more of that anon.)

Perhaps I was doing myself an injustice referring to my lack of education and perhaps I am giving you the impression I have a hang-up about it. In fact it troubles me not, partly because it was my own decision in 1939 to drive tractors and not continue with my education, and as time progressed it became more and more obvious that the experience I had acquired over 25 years in the tractor business was standing me in good stead. In fact I did not work in the Fordson Sales Promotion office very long but enough to realise that although in many skills the people who regularly worked there had me right out of my depth, when the conversation turned to planning or organising a Sales Promotion function, or demonstrations and the use of tractors came up, on so many occasions I could say, 'That will not work but if we try it *this* way it should be OK.' So in a few weeks I was able to do reasonably well in the general office routine; in fact after a while I found even the experienced supervisors would often ask my opinion and so I could feel more and more confident.

My wife Betty and I were expecting a child in January. Betty was living in our cottage in Derbyshire whilst I worked Monday to Friday in Essex, hurrying home on Friday evening for the weekend and returning early Monday morning for another week. Apart from being away from Betty it was not so much work as sheer pleasure, although life never runs quite as smoothly as one might wish. This truth was again proved during the Christmas holiday in 1960. My mother had invited us for Christmas Day dinner, but a great deal of rain had fallen and the result was the usual flood in the village of Ambaston, near Derby, where my parents lived. The street was over 2 ft deep in a raging torrent of water. I tried to drive up it but it soon became clear the water was too deep. Matters could be improved a little by driving along the footpath and at least we managed to get within shouting distance of the house. By this time we realised our Christmas dinner was a lost cause: the flood water had put out the fire in the cooking range and the electricity had gone off. My mother was of course very upset but little could be done, so the Christmas bird had been dismembered and she was frying pieces of it on a primus stove. We felt it silly to get wet through trying to wade across the street and anyway Betty was not sure she could manage it, so having made certain my parents had need of nothing we could provide – their main ambition being to see dry land – we retired from the scene, Mother having assured Betty that the suitcase full of baby clothes which had been sitting in a place of honour in the front room was safe and dry in a bedroom far above the flood.

We now had to find a Christmas dinner. For some reason, possibly due to closer family bonding, I cannot remember any family at that period not being together for Christmas, so there was then less demand for hotels to lay on festivities as they do now; and of course because we were going to my parents we had not brought food into our cottage. (In fact the cottage was not being lived in after the holiday, it being considered Betty would be better staying with my mother as the baby was due in a few days and I had to return to Essex.) We retired from the watery scene and drove towards Derby wondering if any of the small shops might be open. Eventually, like corn in Egypt we found a transport café. Why it was open on Christmas Day we did not ask, but our Christmas dinner for 1960 was bacon, egg and chips.

Some days after, when I was back at work, there was a slight panic as in the fog Betty was transported to a maternity hospital where Nicholas was born. On being notified I hurried home to find a very happy and proud Betty and a wonderful new son. All grandparents were of course delighted as were many family and friends; there was much baby talk and visiting so I quietly disappeared in search of tractors, those things I could understand. However, I spent more money telephoning home now than I had done previously.

The Fordson Tracteuropa demonstration team, which we were preparing for a tour in France, was made up of four Thames Trader trucks and a Thames 15 cwt van fitted with public address system. Various sorts of sales promotion equipment were carried on the trucks, plus a generator for the 12 floodlights and all the other things requiring electricity, such as the projector and screen for the cinema, cutaway tractor, fairy lights and the all-important electric kettle for the essential cup of tea. We had a radio-controlled tractor with a plough, and a 40 ft tower carrying a flashing beacon and announcing to the world the single word 'Fordson'. This was meant to show how Fordson were always looking forward and seeking continually to improve the product. This unit always drew big crowds and of course, most importantly, the press, radio and TV, gave us wonderful publicity coverage.

There was a driving competition intended to get the visitors to our demonstrations into the seat of a Dexta tractor so they could see how easy it was to drive. This enabled us to do a commentary on the event for which there was a prize for the competitor who could complete a driving course without knocking over the obstacles and in the shortest time. The commentary could be slanted towards the main features of the tractor whilst its noise attracted more spectators and competitors who left us their names and addresses to be passed on to the nearest dealers. (As they say, you get nowt for nowt.)

We now set up our headquarters in a workshop at the Ford Mechanised Farming Centre near Boreham in Essex. All the equipment needed cleaning and checking. Our demonstration supervisor, John Prentice, paid periodic visits but he had other things to organise so most of the day-to-day running was left to me. It was understood that for the forthcoming demonstration tour of France I would be team leader with two of the demonstration team I had in Sweden and two more yet to be selected. We cleaned, polished and painted. I managed to get a service course on our BTH cine-projector and some simple instruction on a few of the mysteries of the radio-controlled tractor.

I had been travelling home to Derby at the weekend but now I had no transport, as our new Escort van was needed by Betty as a baby taxi. She rushed around from home to shop, shop to grandparents, then to friends. She had a new lease of life – it's wonderful what a baby can achieve! *He* simply rested in a carry-cot in the van, mostly asleep I am told, a very placid young man and not so different in character now, I guess. The day after he was born I had to go, at Betty's insistence, to buy a pram, it being considered unlucky to have a pram and no baby, and after the sad experience of a few years previously there was definitely going to be no pram in our house before a baby arrived. Now of course a pram could not come quick enough, so I went in search of a baby chariot and with the instructions that it must be a high pram with big wheels. I found just the thing, one with a body that could be easily detached from the wheel section so it and baby could be slid into the rear of the van.

My travel problem of journeying home for weekends was solved by a good friend, Mike Woods, who had joined Ford as a temporary demonstrator, just as I had in 1959, and had been found permanent employment at the same time as myself. Mike is sadly no longer with us, victim of a traffic accident, a tragedy obviously much deeper for his family than for me but nevertheless leaving a sense of loss lingering for a long time. He had bought a very good new car upon our return from Sweden, a Ford Anglia, the model with a rear window sloping backwards. Mike lived near Leyland in Lancashire and so passed our cottage on his way home, dropping me off about 8 pm on Friday night. Then on Sunday night he would return and park outside our cottage about 3 am, crawl into his sleeping bag on the rear seat and stay there whilst around 5 am I left to drive us both back to Boreham. It proved a good exercise and very helpful to me. The 105E, as the Anglia was described by Ford, provided fast and economical transport. Quite often on quiet roads in the early morning speeds approaching 90 mph could be shown on the speedo. I could tell without looking at the clock. At about 82 the car would go very quiet and smooth. (I never knew if this was from the engine reaching perfect balance or if we had simply left the noise behind.) However, I had to lift my foot from the floorboards a little to avoid the possibility, as is often the case with an engine, that if maximum revolutions are reached the engine goes smooth before the ominous clatter and crash of an engine blow-up. But the car would keep on building up speed to around 88 mph even with the accelerator at about three-quarters throttle. It served us well and I still remember Mike with considerable affection.

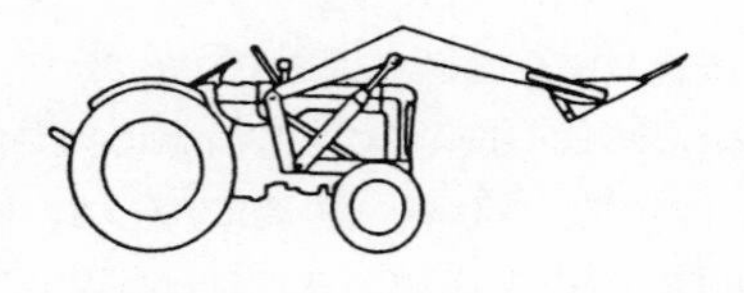

2

IT was 1961. As departure time for France approached we were all looking forward to spring in Paris, as we were to take the Tracteuropa Demonstration Team to the Ford factory and assist with the installation of the Paris Agricultural Show at Porte de Versailles, in those days quite an open space and without the infamous Boulevard Périphérique passing over it as it does today. An extra-long weekend at home and a rather sad goodbye to Betty, but this was qualified a little by the thought of our income increasing with overseas pay and of course the fact that for the next four months I would not be paying for my keep. We had already set our hearts on moving house and living near Boreham, so our being parted and my extra pay were an important part of our ambitions.

The paperwork for customs purposes was quite a pile, as everything had to be detailed precisely and translated into French. A great deal of this work was handled by Thomas Cook at an office in the City of London and during my visits there it was stressed over and over again that we must put nothing on board that was not listed. The day before we left a trainee from the office arrived with 200 Corgi model tractors, saying that Bill Baker, the Show Supervisor, had asked for them to be put on board for sales promotions in Gennevilliers (the Ford France factory). Though it is quite possible Bill had received this request from France at the last minute, I was not slow to warm the telephone wires between Boreham and our office in Dagenham. But I was left in no doubt we had to take them, even if they were not listed on the customs form, so with the threat from me, 'Be it on your head', we put them aboard. Our departure from Boreham was quite low key. I am sure they were pleased to be rid of this team of four who must have seemed to be perpetually in everybody's way and who had now formed a close-knit and humorous team; and anyway perhaps they were a little envious of us touring around France and most of the time out of sight of both supervisors and managers whilst the Boreham staff just carried on running dealer-training courses, good and vital to Ford nevertheless.

We had one driver borrowed from the advertising stores. This was a good way to economise on the Sales Promotion budget because even in those days and even in a company as good as Ford, Dagenham, where the advertising stores were located, was obviously overstaffed and so was subject to all the infamous union pressures that existed in the 1960s. I guess that is why a worker could be borrowed at no cost to Sales Promotion – probably the stores manager either asked for volunteers or used this trip away from Dagenham as some kind of reward for favours past. In this case our borrowed driver was a man called Jim Mitzen, a Londoner living in what today is called Newham. Very humorous, a good worker and excellent company, he had been on the very first Tracteuropa trip in 1959 to Germany so he was pretty well experienced.

So here we were off to Dover for our crossing to Calais. Our first problem was that the rear truck got lost. It is never easy to keep station in a convoy of five vehicles. I always led in the van because it seemed I was better able to read a map spread out on the engine cover (the engine was situated between the driver and passenger seats on these vans) and anyway the van was easier to drive whilst at the same time reading a map than a truck was. Also I was better able to set a speed the trucks could maintain. (It is surprising how many drivers set too high a speed in convoy – it may be OK for the leader to maintain 40 mph but even at that low speed the rear vehicle of five will sometimes have to reach speeds approaching 60 mph to allow for traffic problems if it is to maintain station.) We lost our last truck at some traffic lights on the Old Kent Road. I always watched closely and saw Jim, who was driving, had been stopped at red, so I immediately reduced speed to allow him to catch up. But traffic was heavy and he never made it in the chaos. Four vehicles going slow was causing hooters to blow; and raucous shouts and some V signs from overtaking drivers persuaded me to increase speed. We stopped outside the built-up area feeling sure he must pass along the A2 trunk road; sure enough about 20 minutes later along came Jim quite unconcerned. 'I thought you would be in Calais by now seeing a man about a frog,' he said.

We reached Dover and proceeded undeterred to our contact who would handle the customs paperwork. He had it customs authenticated and signed very quickly and we were soon on board the ferry. When we unloaded in Calais, Cook's local representative, a Mr Dupré was awaiting us. Off to customs and then our troubles started. They only asked to look in one truck and of course sod's law said it had to be the one with the tractors in boxes. We soon had an army of customs officers going over all our vehicles with fine-tooth combs. Mr Dupré was doing his best but had no authority to pay extra duty on the tractors which in any case was quite a lot of money. I telephoned Dagenham complaining bitterly about supervisors who did not know their job, was duly told off for insubordination and told I could not pay the extra duty; the tractors now belonged to Ford France and they must pay. Eventually the office agreed to notify France that we were held up until the duty was paid. The customs would not agree to the models being put into bond so we were shunted into a parking lot and retired to a hotel for the night. The hotel was a good one, I made sure of that. The dining room was excellent, the wine even better. We were a good happy company working on the theory that if Dagenham landed us in this mess they must pay for the problem.

Next day a car arrived with money from Ford France and we were on our way by 11 am. We soon stopped for a two-hour lunch and eventually reached Paris in the evening rush hour traffic. I had realised after a few miles from Calais that we had a problem: the French drivers showing us the way had no idea about convoy driving. It seemed they would drive like mad until we were well out of sight and then wait for us to catch up. I made sure each driver had the correct address of the Ford factory written down and so we entered Paris. Jim was soon lost again but the rest of us, by adopting a French attitude and carving up the French drivers who were intent on carving us up, managed to get to the factory without problems.

One hour later we still did not have Jim. One of our French friends, Bill Marschal, ex rugby player and leader of the French demonstration team, decided to ask the police to help. Another hour and a message came to say the police had located our missing truck parked by the kerb and the driver asleep; they had aroused him and were escorting him to the factory. Jim arrived OK and later I found this was typically in keeping with his character. Give him a job to do, painting equipment, running out cables, in fact anything, you could be sure it would be done and done well. With the job completed Jim would usually simply go to sleep. He once told me, 'All time not spent in sleep is wasted.'

We had a good hotel in Paris and enjoyed the spare time we had in the evenings, visiting monuments, Moulin Rouge, the Louvre, the Champs-Elysées, the Arc de Triomphe and the Place de la Concorde. I ventured into the Louvre one day and was filled with wonder at the magnificent paintings . . . How could anyone acquire such skills and produce such lifelike images? And of course we lost no time in visiting the Eiffel Tower. What an experience going to the first stage was! Far higher than the five- and six-storey houses, in fact one could almost look down the chimneys. But I knew that the tower is 1000 ft high and the first stage is only a little way up. We walked around and took the lift again to the second stage – much higher and beginning to get unnerving especially if, instead of looking down to the ground, one looked up at the lattice framework gracefully curving to the square top floating so far above and overhanging its supports on all sides. Could I really imagine my fear of heights allowing me to go up there? I suppose if one is team leader all things have to be possible, so off we went to the top. It was especially fascinating to peer carefully over the rail and look down at the feet of the structure giving one confidence they were securely rooted in the bedrock of Paris; but even so it was some time before I could force myself to lean on the rail without the fear that too much weight on one side might cause the whole structure to topple over! Of course these were ridiculous feelings but at the top of the tower your senses are different to the ones felt on the ground. The view from the top is not describable to a pen such as mine, but there lies Paris – the Sacré Coeur, Arc de Triomphe, the River Seine with its bridges and barge traffic so far below, even the magnificent street layout can be seen . . . how wonderful!

Bill Baker arrived to give some help and back-up to the

French showstand planners and to enlist our efforts in preparing tractors for display. He only stayed a few days but certainly got things moving, getting tractors released from the factory; and when he found the promised transport did not arrive to carry the tractors to the showground, he telephoned County Commercial Cars Ltd at Fleet in Hampshire and obtained permission to use a six-wheel County conversion of a Thames Trader lorry, which was destined to carry a County crawler around France with us on our demonstration tour. I was partly instructed and partly volunteered to drive this long and rather clumsy vehicle across Paris from the factory at Gennevilliers to the showground at Porte de Versailles with two tractors on the back, returning to the factory empty to collect another two. In truth I was rather overawed by all that traffic but eventually became used to the cars and the mass of other traffic overtaking my rather long vehicle – the Citroën 2 cv cars especially, dodging and leaning over as they swept around the truck almost cutting off the front bumper in their haste to reach an unknown destination. The way most of them drove led me to believe they felt heaven would be their lot in the next life and the sooner they reached it the better.

One particular place was quite tricky, a meeting place of five streets, and priority to the right was the order of the day and jealously protected by the French motorist who intended to keep it like that. However the length and size of my truck and the fact that it carried foreign registration plates probably brought those motorists as near defeat as they had ever been. I eventually became used to gently edging into streams of cars and vans in spite of all the screaming hooters (their use not banned then as it is now). But there was little they could do – I was bigger and, when I became used to the situation, just as aggressive as they were. One day I found a policeman had taken over traffic control. Standing on half a barrel he had traffic flowing quite well, with his piercing whistle and exaggerated arm movements. Alas, later in the morning on my second run, the five streets were blocked. We still had a policeman on a barrel, now red faced and still blowing his whistle, less arm waving now and his jacket removed, cars all around his position not moving, I watched with amusement as exasperated drivers now used their own version of exaggerated arm movements through the car windows with the on-going scream of hooters. (Only an irate French motorist can talk and pass abuse with the use of a car hooter.) The real disaster was that it was lunch time and from what I had already learned of my French friends' habits, lunch time was of much greater importance than the end of the world! The policeman eventually stuck his whistle in his shirt pocket, slung his coat over his shoulder and worked his way through stationary traffic to the pavement. I never knew if he had just given up or if it was his lunch time too.

I learned two valuable lessons from driving that truck daily through Paris, the first was *Patience, but keep on creeping forward* the second *Always tie your load down*. In my ignorance I had not considered it necessary to tie down tractors. After all they were heavy and the truck had side boards on, the first tractor front wheels being set against the lorry front board and the second tractor front wheels being driven up to the rear wheels of the first one – so how could they move? They did not move until I came to the last load of Fordson Majors and there was only one left to carry, all 2½ tons of it. Handbrake hard on and the transmission in bottom gear, I set it over the twin rear axles to give as much traction to the rear wheels as possible, not to stop wheel spin on accelerating (there was never any danger of that happening on that truck) but to ensure as much braking capacity as possible was available so the dashing style of the French motorist could be allowed for, off we went on our journey across Paris. Somewhere on the journey the traffic lights beat me and it became necessary to stop much quicker than I wanted to. The result was a sickening crash from behind and looking in the rear view mirror there was no Major sitting over the twin rear axles but just a Major radiator aggressively filling the view. It had slid from the rear of the flat body right to the front and the only thing stopping it from joining me in the cab was the front board which had now acquired the characteristic curve which became familiar to us throughout our French tour.

When the Dextas were ready for the show I again turned to truck driving. The first two were delivered and with the second two on the lorry I found the gate to the showground blocked. A line of men stopped me, insisting with the word '*Grève*'. (I now understand it meant strike but on that occasion I had to agree it was a most grievous problem.) However, a Ford employee turned up to rescue me, so I was directed to a holding area on an internal showground road high above our showstand, in fact on top of a bank overlooking the lower level. We unloaded the tractors and I carried two more loads so we had six Dextas sitting on top of the bank.

A steep range of steps leading down to the area adjacent to the stand was the only access, with the main gate still blocked by the strikers and the show due to open the next day. The steps were divided into three flights with a landing between each flight so I had a chat with Bill Marschall and persuaded him to join in a venture to drive the Dextas down the steps to the showstand. Most people said it was impossible but I had other ideas.

I have always had the utmost contempt for anyone daft enough to take part in a strike, never having seen them achieve anything of permanent value, so I felt it was necessary to score a small personal victory. The steps were plenty wide enough for our purpose with a bank of flowers and a steel handrail on the right and another steel handrail on the left guarding a drop to ground level . . . easy, I said. There was a rush of volunteers to drive the first one down, but I insisted we must prepare the tractors first. I found a length of timber, similar in size to a step, and let down the tyres until they almost, but not quite, flattened when driven over it. I prepared to try the first step, figuring that even if the tractor got away the steel rail might restrain it from serious damage. With my heart in my mouth and beginning to wonder why the hell I wanted to get involved in someone else's war, I approached the top of the steps, bottom gear, low engine revs. It was a frightening sight to look down what now appeared a very steep set of steps indeed. No one asked Henry Ford if he wished us to risk destroying his tractor but anyway he was in the USA 2000 miles away in his Detroit office.

The Dexta hit the first step, the deflated tyre cushioning the bounce we might have expected from hard tyres, and we gently went down the steps like a high-stepping horse. Perfect! There was a rush of drivers to try it and in no time we had our Dextas on the showstand. I carried more Dextas to the show which were soon in place and polished up ready for the next day's opening.

Bill Baker departed for England and we were on our own in France. We enjoyed the show and the night life of Paris, the Moulin Rouge, the night clubs around Montmartre and of course our long dinners in various restaurants.

After the show ended we helped clear the showstand and then were invited to a rather splendid dinner on the first floor of the Eiffel Tower. This was intended to act as a press function to glean as much publicity as possible for the demonstrations around France during the summer and also to introduce us to French Fordson tractor dealers who would be our hosts at the specific venues we would visit. The Eiffel Tower restaurant was quite a swish set-up so the press, radio and TV people were quite willing to come and enjoy being entertained by Ford France. The dealers were also present and of course so were the English contingent. It had been stressed that the radio-controlled tractor must be present and the plan was to demonstrate it within the four legs of the tower itself; it was hoped to catch the eye of television and thus obtain a good return on the vast amount of money that must have been invested in the meal and function previous to showing the tractor under the tower.

The publicity for this tour was organised by a publicity agent by the name of Charles Thann. He would arrive at each demonstration a day in advance along with a male commentator for the public-address system – a handsome guy and he knew it – also three or four attractive young ladies to staff the tent where the literature for all the machinery was made freely available to the public. The Charles Thann staff dressed in their red and blue uniforms (red and blue being the colours of Esso, who were co-sponsors for the tour) were all present at the Eiffel Tower function.

To obtain the best impact they wished me to drive the radio tractor from inside one of our trucks where I could not easily be seen. I had tried this before and knew the metal frame of the truck sometimes disturbed the radio system, being in such close contact with it. My suggestion was that the tractor should instead be driven from the first stage of the tower. I could look over the side of the balcony and have a splendid view of the tractor in the centre of the four legs below me, the unknown factor being what effect this immense structure of steel might have on our signals. In the event it worked wonderfully well; we tried it during the afternoon, much to the fascination of the general public.

During this rehearsal Bill and I decided to see if we could improve on our performance of driving Dextas down steps and this time drive a Dexta up steps to the first floor of the tower for display in the restaurant. Bill took himself off to the tower office to seek permission which he eventually obtained – not easily, I was assured. (I did not enquire if money had changed hands but just knew Bill did not give up quickly.) We had previously measured the lift to ensure the Dexta would fit into it. Weight was not an important factor because the number of people the lift was rated to carry far outweighed the Dexta. The tractor tyres were set at the requisite pressure and I drove the Dexta up the steps and into the lift. (Am I the only man to have driven a Dexta up the Eiffel Tower? I ask myself.) In no time we had the tractor sitting comfortably in the entrance to the first floor restaurant.

The dinner was long and magnificent – five courses, four wines, coffee and brandy. It is nice to live well sometimes and get paid for it! The speeches were also long and mostly unintelligible to us but we were made to stand up in turn, dressed in our standard white shirts with Ford tie, Ford blazer embroidered with the Fordson wheatsheaf in gold thread on the pocket and grey trousers, as part of the introduction of the English Ford team to the French personnel. Afterwards the press and television assembled at the foot of the tower to see the radio tractor go through its paces. No one thought to look up to the first floor to see the driver, and so afterwards my presence was demanded on ground level just to prove it was not a gigantic hoax. After a few people had been allowed to try their hand at steering it between cones, everyone seemed satisfied and we made TV news that night.

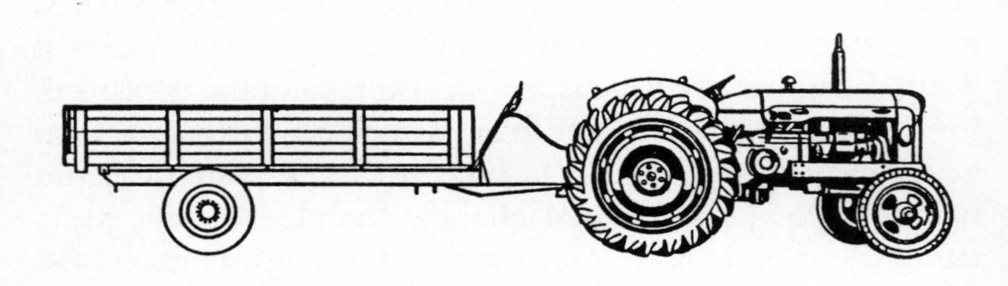

3

WE had various French people to travel around France with us and always several dealers would arrive at the site but our two main support people were Maurice Puget and Noel Tissier, the two Ford France service representatives, Maurice covering the southern half of France and Noel the northern half – two very large areas. Our first demonstration site was at Valence 100 km south of Lyon and a long way from Paris. I was asked if we could take the radio-controlled tractor to an agricultural show in central France for a day while the rest of the team drove to Valence. Noel volunteered to drive the truck if I would follow in the van so he could use the P.A. equipment for commentary purposes. (This all came about because the dealer involved was so impressed with our tractor at the Eiffel Tower that he simply must have it at his show to pull a fast one on his competitors.)

I have forgotten the name of the small town we visited but it was a full day's drive from Paris. Having driven a truck on our Swedish tour the previous year I had decided although the Trader trucks were good to drive and a reasonable amount of power, especially as ours were fitted with two-speed Eaton rear axles, the brakes were lethal; the pedal always felt hard enough and not spongy but often I felt my leg would break and still the truck was not stopping well. It was, I found afterwards, a general characteristic of our trucks. One hill we descended was quite steep with a few corners thrown in for good measure. Following Noel I became quite concerned about the speed but could do nothing about it. Then the smoke started to come from the rear wheels. It was obvious by the smell it was brake-lining smoke, more and more of which came from the rear wheels until they were almost invisible. The smell was like the Ferodo factory burning down. Luckily we made the bottom of the hill safely so I passed Noel and stopped. He climbed down from the cab with his usual miniature cigar between his teeth and we listened to an assortment of crackles and sizzles from the rear wheels. He just shrugged his shoulders. 'It is nozzing,' he said, climbing back into the cab and driving away just as hard as ever.

On reaching our destination we were really feted by the dealer. In common with many French towns this one held its annual fair in the main street. Showstands stood at the side of the street and the centre of the road became a pedestrian precinct. We did several half-hour demonstrations and as usual some people were allowed to steer the tractor using the control box. The dealer might suggest one or two of his important customers, but left to my own devices I usually invited children or, better still, attractive young ladies; but of course my hand was never far from the emergency stop button. It was quite fascinating to see the expressions on the faces of those working the radio, which I am sure made a lasting impression on them of the technical wonders Ford were capable of.

Valence sticks in my mind because of our hotel, not what we were expecting but a large private house, which was or had been something of a religious house. I cannot say we were looked after by nuns, there was no uniform habit but the staff were elderly ladies – we never saw a man – who never spoke unless it was required. They maintained a truly French table and a pristine house with rooms to match, making it a very relaxing stay after the excitement of Paris. Valence was our first full demonstration in France and we soon found out that French farm fairs meant just that, very little demonstration work but at every site a very good driving competition was organised. The winner from each fair competed at our final event at which the winner took home a new Dexta Tractor. Our cinema with French language films was in demand and floodlights gave us the facility of making a really good evening event.

One day much to our surprise two people turned up, both English; one announced he was to drive a Merton dumper around France calling at our demonstrations and doing other demonstrations as he drove from site to site, the other chap, a Yorkshire man who arrived in a British registered Ford Consul estate car, was the Steelfab representative for Europe. He stayed with us most of the time showing mostly tractor front-loaders and promoting stone-pickers, so with four English Ford personnel, one County Commercial representative with a crawler and four-wheel-drive tractor and two gentlemen plus the occasional visitor from Dagenham we had a hell of a social life as well as working and travelling hard.

One of our sites was at Perpignan at the foot of the Pyrenees. It was Easter, only just April, and to me who of course had never been so far south before it was amazing to see peas in pod and potatoes in flower. How wonderful this green garden land seemed with its acres of vines just breaking into leaf! We were invited for an informal lunch one day in the local dealer's workshop. Apparently his service manager had sent his apprentices out the previous day on a snail hunt in the lush hedgerows of the surrounding valley. The snails had been left overnight in slightly salted water (a cleaning solution I suppose but pretty uncomfortable for the snails I guess). The next day a bonfire was lit just outside the workshop door and the snails placed in a circular shallow pan, open end of snail facing the sky, a knob of garlic butter pressed into each shell, and the pan placed over the open fire. There was soon quite an appetising odour, mostly garlic, but by now with the exception of Jim, we had begun to enjoy this most un-English of flavours. A table contained baguettes of bread, a large slab of mountain cheese, and many bottles of wine. After some initial trepidation, I certainly enjoyed my first taste of snails. We simply held the shell in a serviette, stuck a cocktail stick into the shell and winkled out the gristly and rather chewy body with its strong taste of garlic.

Another unusual event occurred a few days later. We had an event to operate (I cease to call them demonstrations, we hardly ever used a tractor in a field) at a holiday centre called Méjanes, near Arles in the Camargue. It was in a vast nature reserve at the mouth of the river Rhône, both windy and wild, with many acres of shallow lakes with many waterfowl of all kinds and, when we were there, great flocks of flamingoes with their flashing feathers showing pink under the wings when they flew. Méjanes had a good restaurant, stables with many horses and a bullring, also small cabins where one could stay overnight. We were not lodged in the cabins because they were reserved for several important visitors who were to attend this event, the Export Sales Manager and the Tractor Manager of Germany (who happened to be an Englishman who had been in the right place at the right time after the war and had made a success of what I suspect may have been a stop-gap appointment but had turned out a long-standing success).

The Camargue is a wonderful place and we were fortunate to be in the care of the proprietor of Méjanes, the owner of which is a company called Ricard, maker of an alcoholic drink flavoured with aniseed, so beloved of the French. Our hosts took us on a trip out into the wilds, in part of the nature reserve normally closed to visitors but we were privileged to take one of the daily trips to inspect the wild white horses and black fighting bulls. These magnificent animals run wild in the vast wilderness and are indeed a wonderful sight to see. The fighting bulls are vicious; they ignore men on horseback or in a jeep but will charge a man on foot and some of the cows are even worse. One cow with a damaged leg was held in a pen in Méjanes and its bull calf of only a few days old would instinctively charge anyone who entered the pen. The bulls, which are not killed in the French bullfights, have a string of rosettes tied between their horns and the bullfighters, uniformed of course, are rewarded for each rosette they can steal but like Spanish bullfights they are only allowed to attack the bull as it charges at them and must not move against it whilst it is standing.

With our equipment set up I was amazed to see an English Leeford ditcher arriving on site and, yes, it was operated by a man whom I had worked with in Ireland previously and also at agricultural shows in England. It is amazing how as one visits shows, etc, the same faces keep popping up. The dealer involved in this event invited his customers to come on their Fordson tractors. About 60 turned up and each one had a voucher to spend in the dealer's shop.

Both afternoons of the two-day event there was a bullfight. Beforehand we had a tractor parade and a demonstration by our radio-controlled tractor in the bullring with, of course, our film-star-like announcer on the P.A. system and the lovely publicity girls to hand out leaflets. The bullring was filled; in fact I was told 15,000 people had visited us, causing a traffic tailback of four miles. This was the first bullfight I had seen. Later the Spanish ones were to fascinate me, but this was a normal French one. I was determined to get some cine film of it and asked if I could go down into the passage around the ring where the bullfighters took refuge between attempts to claim a rosette from between the bull's horns. This was a desperate mistake! After the first bull had his rosettes stolen by the bullfighters (who incidentally wear white shirts and flannels with running pumps, no armour against a bull's horns except their own speed and agility) there came into the arena a particularly vicious and fast bull. Eventually one of the bullfighters, determined to take a rosette, got really close to the bull and managed to snatch one; but being so close he had no sort of start on the bull, who chased him to the 5 ft barrier around the ring. As the agile bullfighter leapt for his life the bull also leapt and was in the passage with the bullfighters and me. Luckily I was on the far side of the ring. As the bull charged around the passage scattering bullfighters over the barrier and into the ring, a gate was opened which blocked the passage and diverted him back into the ring. The bullfighters promptly went back over the barrier, leaving a pawing and snorting bull in charge of the ring.

Eventually the bull had his three rosettes all stolen but he was still pawing the ground and threatening anyone with death and destruction who tried to drive him out. Then a cow was let into the ring, I suppose to calm him and lead him out. But someone had made a sad mistake – this lean young cow, ribs rippling under its black hide, was I am sure a 'racing cow' if there is such a thing. It was faster than the bull, and with horns curving forwards and upwards seemed to me more dangerous and better armed than a Sherman tank. Something had to be done. The bullfighters all crowded into the ring waving shirts, cloths, towels, anything to distract the two animals and work them towards the exit, but the pair had been there before and no way were they going out.

Meanwhile I was filming and enjoying the spectacle until a bullfighter got too close to the bull again and we had a repeat performance of the bullfighter leaping the barrier with the bull following, but now it was a different ballgame: there was a bull in the barrier passage and a fiendish cow racing round the ring. Was it worse to be crushed by the bull or spiked by the cow? If I believed cows to be gentle creatures with good maternal instincts, I was wrong – this cow went for anything that moved. As the bull approached I went over the barrier, as the cow ran across the ring I went back behind the barrier. Eventually a man on a horse came into the ring and with his three-pronged *guardienne* stick managed to get them out of the ring. I did not stay in the passage to take any more film but enjoyed another similar spectacle from the safety of the grandstand.

You may well ask what is all this to do with demonstrating tractors. I often wondered myself but really it was a Ford France show; we were there to give support with our show equipment and remember there were some machines being demonstrated, as well as a driving competition making headlines in the local papers with speculation whether a local driver could beat all comers at the final show and win the Dexta tractor, and a cinema plugging Ford for all it was worth. There were lots of new tractors on view and many of the sites were unsuitable for tractors at work. Here at Méjanes for example there was no agricultural land; other sites were in town squares or in streets or aerodrome runways. (Can you imagine the local pilots' reaction to someone drawing a furrow across their grass flying strip?)

I remember another story also involving an aerodrome. This time it was at Pontivy in Brittany, where our hosts were members of the local flying club. It was hard work from day one as apart from setting out and painting our equipment we were being right royally entertained in the flying-club bar. I believe the instigator of this entertainment received by both the English and French teams was Bill Marschall who, before his employment by Ford, had been involved with the public relations side of the Tour de France cycle race. That and his rugby background gave me the impression that he (and of course we) were revisiting places he had been to before. This event at Pontivy definitely carried echoes of previous encounters where great orgies of food and drink had been lived through at some time. Lunch was at 12 noon in the flying-club bar and afterwards in the excellent restaurant. Most days we had a guest, a local judge or the town mayor, perhaps an important landowner, and so lunch usually lasted until around four in the afternoon with lots of food, wine, and rugby-type songs. Afterwards we were poured out of the bar and attempted to carry on with our preparations.

Eventually the day of the event arrived and so did people. We had 20,000 visitors and our formerly grand lunches were now replaced by a hot dog and coffee if we had time or a cup of tea from our electric kettle powered by the Ford 100 HP six-cylinder engine and generator. At the end of this hectic day Bill announced that there would be a Ford dinner in honour of the flying-club chief. He was an aerobatic champion and had been part of the French national aerobatic team a year or two previously. He certainly was brilliant. No Pitts Special plane for him: he gave us wonderful displays with an old biplane of wood and canvas construction. Everything he did was different to the flying I had seen before. During our day's event he did two displays. I was particularly taken with his original manner of looping the loop. Unlike most fliers, as he looped he was on the outside of the loop and not on the inside. He would then climb the plane vertically until one could hear the engine almost stalling and just at the critical moment when we expected the plane to fall backwards out of the sky he would turn it completely around and glide back down the track he had so laboriously climbed a few seconds before. Another of his favourite manoeuvres was to approach the demonstration site low on the horizon and at high speed (if that term is applicable to his old biplane); just as he reached the edge of the field he would flick the plane over on its back and with a wave to the crowd would roar over their heads about 20 ft high. The downside to this was that the plane did not have a pressurised fuel system and with the fuel tank now underneath the engine there was soon a heavy misfire and the engine died. Quickly flipping the plane right side up, he had everything timed precisely because as the plane was righted the engine came back a second or so later and he soared off into the sky. I often wonder if he lived to draw a retirement pension. We all called him 'the ace' – it came naturally.

That night's dinner, organised by Bill with the ace taking the place of honour, was quite something. The food was typical of the region, shellfish, seafood of all kinds, dry white wine with cognac or calvados to follow with coffee. When the ace decreed we must drink champagne, we needed no encouragement at all and after many stories and songs found ourselves, at 2 am, watching Bill and the ace beating out a rhythm on the table with an empty wine bottle in each hand whilst one of the publicity girls danced on the table. Most people joined in a light relief using cutlery. Eventually the dancing lady decided enough was enough and fell onto the ace, who, carefully removing her shoe and making sure there were no holes, filled it with champagne to revive her. Afterwards the lady went around the table so we could all partake of this very special champagne. It was a memorable evening.

The next day we were on duty at 10 am. For the first hour or so I am certain some of the flair of our show was missing. Even the ace was subdued . . . no outside loop on this first show. When he came down I laughingly asked what was the matter. 'Zere is zumzing rong in ze ead,' he replied.

About this time France was experiencing considerable political problems in Algeria, which was agitating for independence. At our next location we lived in a hotel on the main square in the city of Nantes and were amazed to find a policeman with a submachine gun barring our way as we returned to our hotel one evening. After inspecting our van and looking at our passports he allowed us to pass. At the hotel we were told they were expecting a big demonstration against Algerian independence in the square and were advised to stay indoors. The next day sure enough there was a milling throng in the square. It was probably May Day and a holiday in France, but at that time this was not an official bank holiday in England so I could telephone Dagenham and explain what a hard time we were having; by holding the phone by the open window, I made our office clearly hear the chant of thousands of voices crying 'Algerie française' accompanied by the rhythm repeated on motor-car horns. The square filled with people and police. I certainly did not intend to leave the hotel, but listened to excited speakers whipping up the crowd to a patriotic frenzy. On the outskirts of the crowd there was a hooligan element which soon started to break windows, our hotel not excepted. The police took immediate and severe action; the hooligans were roughly handled and soon sped away in black vans.

After our more pacific tractor demonstration, we departed for our next site at Bourges. I was feeling very much under the weather and after about two hours' driving could go on no longer. A doctor was called and he diagnosed severe shellfish poisoning, telling me not to eat for several days and to drink Vichy water only. He gave me a foul-tasting potion to drink, but as Granny used to say, 'If it isn't nasty it is no good,' so I persevered with the foul stuff. Bill Marschall came to my rescue and took me in his car to our next hotel and so I retired to bed. In all I was there for a week. It is indeed a lost week to me: all I can remember of it is occasionally awaking and ringing for more Vichy water. I tell you this so the story can be balanced. There is always a down side to life and if you enjoy the good things always be prepared for the bad side to hit back just when you either do not expect it or can definitely do without it. I often wonder how ill I really was because it certainly felt awful.

Just before Whitsuntide we were working in the city of Reims. I had made arrangements for Betty and Nicholas to fly to France for a holiday. Jim had also arranged for his wife to visit so Jim and I set off for Paris to collect our wives from Le Bourget. It was wonderful to see Betty walking down the steps of a Vanguard airliner, with Nicholas being carried by a stewardess – in the 1960s passenger care seemed so much more important than it does today. On our way back to Reims we stopped for strawberries and cream at a roadside restaurant and even today, so many years later, we still occasionally remark on the size and flavour of those berries. Memories are indeed a wonderful thing! Or was our joy at being once again together giving us a rose-tinted view of life?

After one night in Reims and its inevitable good French meal we were to travel to Dijon for the next show, so off we went with the team's suitcases in the van, Jim's wife sharing his truck cab and Betty and Nicholas in the van with me. (Nicholas had, I understood, been the cause of some friction back home, grandmothers being absolutely certain that babies of only a few weeks of age should not fly in aeroplanes, never mind being dragged around a foreign country in a van. However, it all subsided and he survived trains, aeroplanes, French hotels, tractor rides, demonstrations in the sun and dust, all without any effect at all . . . tough old things are babies at 16 weeks old.)

The radio tractor had been giving some problems at the event in Reims so I had arranged for the dealer in Dijon to locate a radio engineer to look over our prize exhibit as soon as we arrived in the city. During our journey the trucks left Betty and me on a layby to feed Nicholas and so we were driving alone as we approached Dijon and were signalled to stop by a policeman waving a firearm at us. Betty was most concerned, I could say frightened, that in a foreign country, so far from home she should feel threatened by a gun. However the police only needed to look at our passports and check the contents of the van. I will never know what they thought of an English man and wife travelling together with 14 suitcases – I guess they must have thought we were either mad or very rich. Apparently the police were checking everything on wheels because President Kennedy of the USA was shortly due to pass that way.

We had been booked by Ford France into one of the

best hotels in town, La Cloche, the sheer luxury simply overpowering Betty and Jim's wife. Dijon calls itself the gastronomic capital of France so you can guess we lived and ate well at La Cloche. But this good living could not be sustained; it was not possible to lose the costs of our wives living at those high rates so we had to find a lower-price hotel where we could obtain a double room for the cost of a single one at La Cloche and our finances were back under control again. Even so this hotel was quite good; we even had four taps in our bedroom: hot and cold water with two more marked red and white wine, along with the appropriate glasses.

The first day's work saw our radio engineer arrive to check our equipment. By this time our tractor had lost several functions, the worst problem being that it only steered to the right. Half a day's work and a few hundred francs later the diagnosis was that it needed English spare parts that the engineer did not have. There was panic in the camp. The radio tractor was the mainstay of our press, radio and TV campaign and something had to be done. I telephoned the manufacturer in Weybridge who said it would be no problem to repair it in their workshop. What to do? We had no separate customs paperwork for the radio part but I decided, with the backing of Ford France, to take the equipment to England and Ford arranged for a customs expert from the Ford factory to meet me at the station in Paris. Since all the documents we had were signed by me it had to be me who took the equipment home . . . So after two months of being parted from Betty here she was in France and I was returning to England.

A first-class ticket was booked on the crack express of those days, 'The Mistral' travelling from Nice to Paris. It arrived 20 minutes late in Dijon for the two-hour-and-fifteen-minute journey to Paris but two hours and five minutes later we arrived at the station in the capital. Two hundred miles from Dijon – some train, some speed at least in those days! The Ford man met me at the station and off we went to Le Bourget. A certificate of export was obtained and I flew to Heathrow, where, putting my suitcase on the customs counter, I asked for the officer's assistance. 'There is not much I can do officially,' he said, 'but I will stamp your French certificate with the import time and if you come back here tomorrow night, I am on duty from 10 pm onwards. Ask for me.' I booked a flight back to Paris for the following midnight and took a taxi at vast expense to Weybridge. Between 2 pm and 8 pm the next day the equipment was repaired and I was driven back to the airport. Now would I get out of England and indeed would I get into France again? There was a lot of hanging on the next few hours. Everything went like clockwork: the customs officer stamped my certificate, the French customs looked at it and waved me through, and the next afternoon I was back in Dijon. Not only was I reunited with my family but we had a tractor that worked perfectly.

After two weeks in Dijon we reached Clermont-Ferrand, where at last we were able to do some ploughing with a Dexta. This demonstration was notable not only for having working tractors but for a magnificent luncheon complete with a society of French horn players and a sucking pig roasted and beautifully garnished. Betty and I had seen nothing like it. By now you might be excused for thinking we are not on a tractor demonstration tour at all but a gastronomic tour of France; but do remember that although we were living well and being treated as VIPs, particularly myself, the days were long and the site work hard. Many a day I would have loved to return to my hotel at perhaps 6 or 7 pm instead of 10 or 11 pm as was usually the case and simply gone to bed without any dinner at all, especially when I had Betty for company, but alas duty was always calling to meet this press man or that important customer, or maybe the French tractor manager would be calling to see how we were and receive Bill and myself for dinner.

About this time Ford decided that the Fordson tractor offices should move from the Dagenham factory to new ones on the top floor of the C&A building in Ilford. I did not realise it but this was the start of the American influence in the company becoming dominant. I do not think many people in the company realised this, but obviously if I am correct in this theory top management did know. In America Henry Ford the second (or as he was nicknamed 'Hank the Deuce'), who had taken over from his grandfather some years before, had been through a torrid time when the company was losing a million dollars each day and had by now managed to turn this massive loss into a small profit; but he came to believe that Ford must become a worldwide company as it had been before the war, if it was to survive the last part of the twentieth century. The result of this was that the American company now began to gain financial control in several overseas Ford companies in such a way that it could play a much stronger role in the decision-making process of these companies. I believe that changes in the English company were part of this process; our freedom to trade was, I believe, becoming more and more controlled from Detroit. Perhaps someone in Detroit had decided 'Those goddamn tractors take up too much space in Dagenham,' and our office was moved to Ilford, pending a further move to a new factory to be built in Basildon, so more and more cars could be produced in Dagenham. After all, cars were the traditional profit-earner for Ford.

We still had Mr Batty as our tractor manager. He terrified almost everyone – a real manager – no excuse was acceptable: either you did what was required or you moved on. But he was scrupulously fair, a characteristic which must have been recognised in America, because he eventually became UK General Manager. No one deserved it more. And how in subsequent years he would be missed in 'tractors'!

One situation brings his character into sharp focus for me. It was after my return from France and during a stint in the office in Ilford. Apparently there was a service problem with the Fordson Major, to do with resiting the dash panel; the isolator switch now needed an extra two inches of wire to connect it to the wiring loom. It seems to me that the purchase department did not want to instigate a request for a wiring loom with an extra two inches of wire in it because this would have meant renegotiating the price of the loom with the supplier and inevitably increasing the cost of the loom by a few pence, which would have been frowned upon by the 'bean counters' (financial department) as we called them. Manufacturing were stretching the wire until it just reached but it was really tight, in the field it was becoming detached and if it touched metal the whole of the dynamo output would go to earth, causing the loom to catch fire. This caused many warranty claims, culminating one day when a dealer who had supplied several new tractors to his best customer found to his horror that three of them were out of commission due to burnt wiring looms. Worse still, wiring looms were not available from the parts operation. This resulted in a fiery telephone call at 10 am to Mr Batty. At 10.05 Mr Lasman the service manager, the parts manager and I were in Mr Batty's office. The service manager confirmed there was a warranty problem and the parts manager confirmed that wiring looms were out of stock. Mr Batty picked up his telephone and instructed the car pool to make available a car within ten minutes for me to collect. Production at Dagenham were instructed to prepare six wiring looms with an extra two inches of wire installed for me to collect in half an hour. By 10.30 the dealer knew I was delivering the wiring looms to him that afternoon and the bean counters were told to make sure this situation was remedied that day. Now that is what management is all about.

Our final demonstration in France was at Arras in the north. Crossing Paris I dropped Betty and Nicholas off at Le Bourget for their flight home. A postal packet arrived from Ilford, quite a bulky one telling me to take the demonstration equipment, not home as I expected, but to Cologne, where I was to leave it with Ford Germany before returning for our holiday. As ever, another Ford surprise, but they had sent another set of documents for our tackle, this time in German. The demonstration at Arras was important because it was the culmination of all the driving competitions we had held throughout France and someone at this event was going to win the Fordson Dexta. Here in the north we had a field of tractors working as well as our usual equipment in use. The twelve driving-competition finalists were paraded through the town with the P.A. van playing French accordion music, each competitor proudly carrying his name and town on a board strapped to the tractor bonnet. In addition many Fordson users joined the parade with their tractors and so general chaos was produced in the fine old French market town.

4

RETURNING home was wonderful after four months away. There was money available, overseas pay was good and our current accommodation in Derbyshire was cheap so Betty had saved quite a pile of money whilst I had been away and there was now the chance of another overseas trip of fair length to Germany after the holiday. Little wonder then our thoughts turned to a home of our own in Essex. During our holiday we visited Essex and looked around for a suitable house. A new three-bedroom detached house with integral garage was selected, in a village called Heybridge near Maldon, not so far from Boreham. But could we afford it? The price was an astronomical £1999. . . . Would we ever live long enough to pay for it? First-time buyer meant no delay and within three weeks we had a mortgage and were moving in. I actually lived in our new house two days before flying off to Germany for three months, leaving Betty in a strange place in a strange house and with strangers for neighbours. As ever, responsibility comes suddenly to the young. Anyway we considered ourselves lucky. Betty had a new house and a family at last and I took up residence in a rather swish hotel in Cologne, 'The Dom' near to the great cathedral, the very cathedral that all the might of RAF and American bombers had not destroyed during the war although it was seriously damaged. (I formed the opinion that it was too big and heavy even for those formidable forces to demolish.)

With one companion I moved to Hotel Strang in Euskirchen and took all our vehicles out there, where we could paint, repair and generally attend to them and the equipment, over a three week period. One day much to our surprise we found English-speaking people in the hotel, in fact from Fetterangus in Scotland and representing Greys, an agricultural machinery firm which is still making good machinery even as I write this so many years later. Alas one of the party, Eddie Grey, is no longer with us but I can remember with pleasure the week or so these good Scots brightened our lives. Another event remembered with pleasure was a visit to the Nurburgring for an important sports-car race. The previous week we had visited this famous race circuit to buy tickets for the grandstand and found to our surprise that we could drive around the course for the small sum of five marks. Needless to say our demonstration van was soon on the course, which in those days was the big course – all 14 miles of it. My memories now revolve around one corner called the carousel, part banked and quite sharp. I decided it could be taken at 60 mph in a Thames van. It was possible but only just; the suspension bottomed when we hit the banking and many things in the back of the van decided they must go straight on. Of course the side of the van restrained them but the noise and consternation was quite something for a few seconds.

Back in Euskirchen the rest of our team soon arrived from England and we departed for our first site in Mönchengladbach in the north. How well I remember the wind and rain, not to mention the mud! Our main disaster was our cinema tent which, being inflatable, was severely damaged by the wind causing a tear almost a foot long in one of the main air passages. This was one of the worst experiences of all the tours I was involved in, just sitting in the mud with an umbrella held over me whilst I attempted to repair this gash with solution and patching. But it was done and the cinema was back on show during the afternoon. We were demonstrating tractors at every show and drawing the German farmers in their thousands, just as if Ford Germany had thousands of tractors to sell. No public-relations spectacles here, just good big working tractor demonstrations. The tractors were set out in lines, each on its own plot, just as we would do in England. Each machine would make its pass across the field and back to the accompaniment of a commentary from one of the German personnel. It was amusing to see middle-aged and very fit German farmers with highly polished black leather knee boots, riding breeches in green cord, military cut green jacket and green hat with a complementary feather, marching with stiff straight backs across the field in twos or threes. I wondered what those fellows had done in the war.

The nearest we ever got to a publicity event was the firework display that opened each show – wonderfully done, huge bangs and fiery stars in the air with parachutes descending to earth with model tractors hanging from them. As we trailed around Germany, just as in France, there were memorable moments, but not the same sheer crazy humour, the German character being much more serious. But we did our best to enjoy life. We still had a County Commercial man with us and the Steel Fabricators rep with his Consul estate, so we had good company for the evening dinners and visits to various bars around town. We had a memorable dinner in Graf Zeppelin's restaurant in Schweinfurt and visited the Roman remains in Trier. We spent a day in the home of a dealer near the East German frontier who took us along the main road signed to Berlin until we came to the river Elbe. The bridge over the river had been demolished on the East side so the main road from Hamburg to Berlin stopped there, although the West German authorities still maintained the road up to the middle of the bridge. As we stood on the bridge we could clearly see the high wire border fence and the tall lookout posts above it. We saw a man come through the fence to fetch the cows from the meadows leading down to the river and immediately a police launch appeared cruising along the East German side with the East German flag, a circle with the hammer and sickle imposed in the centre of it flying from its stern.

We worked our way through Germany until we came to Munich, just in time for the Munich Oktoberfest. Our two German friends who spoke such good English took us to a large park and into the largest tent I have ever seen, just row upon row of trestle tables and chairs; on one side was a bandstand with a stalwart German band playing itself into a beery delirium whilst strapping waitresses carried steins of beer to all quarters of the tent and chefs provided cooked chickens which arrived on paper napkins to be dismembered with one's fingers. The beer delivery system fascinated me. There was a stainless steel pipe about forty feet long with taps every two feet or so, connected to a round cylinder with a hose. After watching for a time I realised the cylinder was the rear of a road tanker and as one was emptied another would reverse into place. There was no choice of beer, just heavy steins of about two litres. It had to be seen to be believed how many of these large and heavy mugs the ladies could carry and also how quickly a tanker could be emptied. The Ford men decided to buy the band a round of beer. This meant that by tradition someone from our party had to conduct the band. I was set up, wasn't I? In revolt I thought I would see how a bit of English patriotic music went down with the Germans, so here was the old enemy of Germany conducting the band in front of thousands of them in 'It's a Long Way to Tipperary', left hand trying to keep the beat constant and the right one trying to control the volume. I did not do it very well but got a round of applause anyway.

On returning to England at the conclusion of our tour I was urgently summoned to the new offices and told to get one of the trucks unloaded and sent to Dagenham for three tons of wheel weights, bolts, and washers. It was the date of the British National Ploughing Match and I was to run our tractor loan scheme. I was given the file with all the forms competitors had sent in telling us what equipment they required so when the tractors arrived on Thursday I could allocate each tractor to a competitor and get the specification built up to suit him. So I was back to ploughing matches, at which I had spent so many years competing and made so many friends. To say I was pleased would be an understatement. To add to my pleasure, the names of many of my friends were on the lists I had been given. And so as we altered wheel width, fitted wheel weights, cleaned tractors, labelled each unit with a competitor's name, I was as happy as a cow in clover.

5

NOVEMBER and December of 1961 saw me with a desk in Ilford, enjoying a totally different type of life, doing various promotion jobs and gradually preparing for two farming fairs to be held in the spring, at the same time occasionally explaining to various junior bean counters why it took so much money to tour France and Germany. They could certainly add up quicker than I could but were badly lacking in imagination when explanations were called for to justify money used abroad. Eventually they decided they owed me quite a lot more than I had taken with me, as I must have spent some out of my own pocket, so Betty and I toasted the bean counters with a bottle of French wine.

After Christmas the farming fairs were fast approaching, but this time there was no Ford France and no Ford Germany to do all of the preparation work. John Prentice had the day-to-day responsibility and much of it was passed down to me. I always enjoyed working with John; he had original ideas and with having so much more experience than I, a great deal could be learned from him. A dealer committee would be formed, a site found and many other manufacturers of equipment (not competitive to Ford) were invited to take part, so although Ford contributed money to the events budget and also of course staff and equipment, dealers and other participants also made contributions. Indeed those farming fairs were really an agricultural show in their own right, plus a large tractor demonstration. John or I attended most of the dealer planning meetings. Our first event was to be staged in the Mendips just south of Bristol in the Cheddar Gorge area, so for the first time I saw this famous tourist attraction. The second event some weeks later was near Kilmarnock in Scotland in a place called Peace and Plenty so there was lots of travelling to do. I enjoyed it all and gradually I could do more and more of the organising and less and less of the field work.

As I became more of an organiser and overseer than a demonstrator, another character, Jim Taylor, worked on setting up the demonstration site. Jim had joined Ford as I had and if anyone had said that Jim was a better demonstrator than me I might have disputed that, but two years of being team leader and hardly putting a spanner on a tractor unit left me with the feeling Jim was sound in practice and a better man to get each unit set up and working to its best. He was a perfectionist and gave some staff, particularly trainees, a hard but fair grooming in tractor usage.

The show in the Mendips went off very well. Although on both show mornings there was a light covering of snow, it did not put too many farmers off – after all, entrance to the show was free. Possibly late March was not the best time to venture onto the hills but demonstrations have to be staged when dealers and manufacturers can spare time and staff; also of course there is the important matter of 200 or so acres of land to find.

Then the mammoth job of moving everything to Peace and Plenty in Kilmarnock got under way. By now we were getting stretched for staff; a public announcement and unveiling of the Super Dexta was to be made at Boreham, so whilst John Prentice my supervisor applied himself to that, I was pretty well left to get on with the farming fair. Because we were tight for labour two dealer trainees (i.e. dealers' sons or family members who were spending time with Ford) were sent to me to help and see how a Ford demonstration was organised. One of these, Carlos, the son of a large dealer in Colombia, was quite a character. Apparently on his arrival at London Airport from South America the police had immediately confiscated his 6-inch, razor-sharp knife and his revolver, which meant that for the first time since starting school he was unarmed and in a strange country. It had taken some time before he was confident enough to walk around unarmed, but by the time we had him in Scotland he had become accustomed to this. A dark placid character with Spanish good looks, he had acquired a small following of young ladies. He must have spent a fortune on the phone to London: he would talk for hours in the evening.

His main achievement with us was after the event. We were packing up and in a hurry to return south we borrowed a tractor and loader from the farmer to help finish our loading. In my hurry to get away, instead of taking the unit back myself I sent Carlos with it. He came back with an irate caravan owner who now had five nice round holes in the side of his caravan where Carlos had stuck the forks of the muck loader through it. My remark

that it was now air conditioned did not seem to be well received, but there was little I could do at that time. Eventually I had a file two inches thick in my office desk about this mishap. I had to wade through reams of forms but just before I left the UK on our next foreign tour I managed to clear all the red tape and the caravan owner was paid for the damage. It had been most difficult to sort out. The bean counters did not see why Ford should pay so the insurance company were contacted; they declined because the tractor did not belong to Ford, nor was the driver a Ford employee, nor was it Ford land and anyway the driver did not have a driving licence. Me, I just wished I had had the caravan repaired in Kilmarnock and claimed the repair on my expenses. Carlos was quite unmoved; I obtained a statement from him and never saw him again.

Tales gradually filtered back to me about Carlos and his amorous exploits in London. One particular story concerning a car he purchased amused me. Carlos had authority to draw certain monies from his father's account with Ford. Now it so happened that Carlos wanted a car – possibly one of his current girls had moved out of town and he felt the need to visit her or maybe he just wanted to show off a bit. Anyway, he asked for an amount of money so he could buy a Ford Anglia from a dealer. His request was initialled by export sales and Carlos presented it to the finance department, where it was approved and the bean counters gave him the money and he bought the car. His father was made aware that Carlos had drawn this money and immediately wired back to say this exceeded the amount Carlos was authorised to draw and that he held Ford responsible for this blunder and would not agree that it be taken from his account. The bean counters went into shock and, summoning Carlos to the office, told him he must sell the car and return the money to

Ford. With an air of injured innocence which I am sure I would have recognised he said he already had sold the car. 'Good. Please return the money to us then.' 'But I spent it in Soho last week,' said Carlos. I cannot say how this finished but Carlos, for all his air of innocence and simplicity, is the only man I know who had Ford at the wrong end of a deal.

The next Tracteuropa tour, due to start in August, was to Spain, so most of the summer was taken up with preparations. The equipment needed considerable attention and so did the paperwork; again there were many trips into London to make and hours spent in Thomas Cook's office. I spent as much time at home as possible with Betty and Nicholas – especially Nicholas, because when I returned home from the previous tour of Germany this nine-month-old boy had looked in amazement at this strange man invading his home and taking his mother's attention away from him. We hoped this could be avoided in future. At the beginning of July we had a two-week holiday touring France and Germany in our Ford Anglia van, a wonderful trip.

The usual four trucks and my demonstration van were involved. Budgets were running low so I only had two of the original team members with me. The other drivers were Mike Woods, who would only drive to Barcelona and then return to England, a member of export sales who spoke good Spanish but had never driven a truck before, and a dealer's son who was not even old enough to drive a truck; so with this somewhat inexperienced crew we left Boreham for Barcelona at the end of July. My experience in France was very useful: the road to Spain was remembered very well, especially the good hotels and restaurants, we took it quite leisurely and in four days arrived at the Spanish border, made contact with the Thomas Cook representative and in 30 minutes our carefully prepared paperwork had us on the road in Spain. I was now on new territory so picking out what appeared to be a good restaurant we stopped for lunch. An hour later two men approached us introducing themselves, in good English, as being from Motor Iberica. They had come to arrange customs clearance at the border, and found great difficulty in believing we had cleared the Spanish customs in 30 minutes. However, they escorted us to the factory in Barcelona and we were taken to the Avenida Palace hotel in the centre of the city. A five-star hotel – this really was living! We stayed there for several days of sheer elegance.

Next day we returned to the factory and I was taken to see Señor Fabregas, the top man in Motor Iberica, who quizzed me for most of the morning about tractor production in the UK and sought information on any new models that Ford might produce in the future. Later that day we met the Spanish personnel we were to tour around Spain with. We were allocated a driver, Ramon, for the truck now vacated by Mike Woods; always willing and cheerful, he became a firm favourite with us.

Now we should perhaps consider the political implications of our tour. Nothing is simple and touring foreign countries on what appears to be a straightforward tractor demonstration tour always seemed to work out most complicated; and this tour seemed to me more complicated than the others I had been on. Certainly we were to demonstrate tractors but again tractors were easy to sell at that time even without demonstrating them, so why were we here at vast cost to all concerned? I began to believe there were two main reasons.

The first, and probably most important, was that the Ford name should be promoted as strongly as possible, because Ford had no presence in Spain at this time. In the early days of the General Franco administration Ford Spain had been sold either because Ford felt it was not a viable market or more likely because the Franco régime did not want foreign companies involved in Spain; so a private company, Motor Iberica, had been set up to purchase the old Ford company. Motor Iberica were the only importers of Ford trucks and tractors into Spain at this time; they also manufactured a truck and a tractor in Barcelona. The truck was a lookalike of the old cost-cutter truck no longer made in Dagenham and the tractor was a lookalike of the Fordson Major, both being sold under the name Ebro. Whilst we were demonstrating tractors under the name Ebro there was also the name Fordson to promote, so the few Dagenham built Fordsons that could be imported were also promoted.

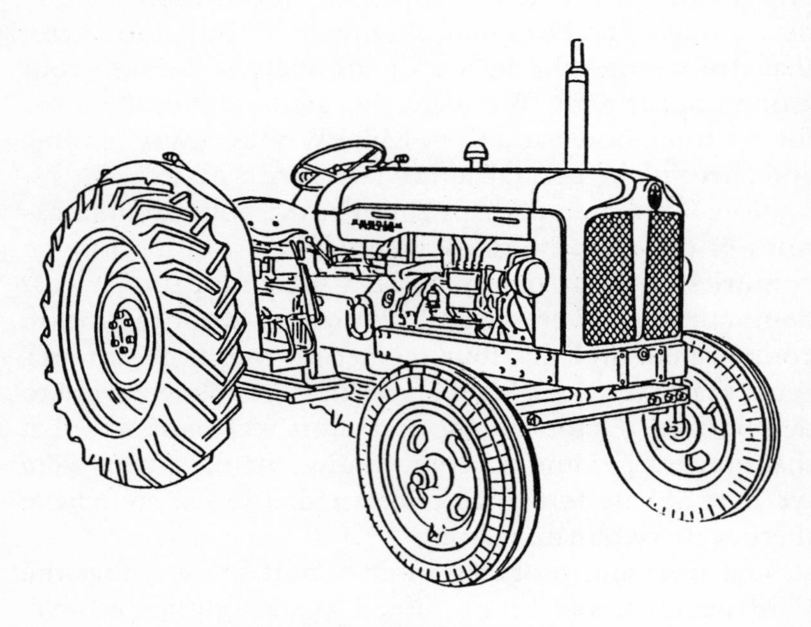

A Fordson Ebro

The second and main reason in my opinion was that this high-profile tour gave Motor Iberica the opportunity to offer entertainment to important customers but even more important the local and in some cases national politicians could be feted and entertained to the advantage of the Spanish company under the banner of the giant international company of Ford, thus perhaps procuring many more import licences to Motor Iberica. As is usually the case, many Spanish users were convinced that Ford tractors were better than the Ebro tractors so it was important that more Ford tractors were available to Motor Iberica which could be sold at a premium. I must say I could see very little difference in the two models.

Later when Ford did eventually re-enter the Spanish market they did so under a different banner so were in competition with Motor Iberica still selling Ebro tractors. This situation was made even more ridiculous when Massey Ferguson bought Motor Iberica and changed all the bonnets and tinware on the tractors but still had one tractor being sold as Ebro with Massey Ferguson appearance but pure Fordson Major underneath. This eventually stopped and Motor Iberica was sold to Nissan, who carried on manufacturing Ebro tractors.

The old cost-cutter trucks being used on this tour were overloaded, underpowered and operating on Spanish main roads that were little better than our country lanes, so when we departed Barcelona for our destination Albacete slow progress was made. We soon had to climb quite a steep hill and the Spanish trucks boiled and boiled and were so slow we had great difficulty in climbing the hill with our Thames Traders because we could not travel slowly enough. Eventually we had to pass them and await their arrival at the summit. We progressed around Spain like this but without complaining because the Spanish themselves were always cheerful and friendly, probably more so than the locals in any country I had yet visited.

Our first site at Albacete was high and hot. We were putting up our equipment in temperatures exceeding 100°F – and this only a few days after leaving England in the usual hit-and-miss summer. During these preparations we were so exhausted by 11 am we just gave up and returned to our accommodation. I have never been so hot and tired, just dropping into bed as soon as we arrived home; never mind lunch – no one could eat any; out to the field again at 4.30 pm and work until dark. I can well understand why they have a long siesta and eat dinner between 10 pm and midnight.

Spain fascinated me. I could see there an image of how England must have been a long long time ago, for Spain was still under almost a feudal system. During the demonstration I was asked to escort the landowner and his son around the site explaining our equipment to them. The father was a real Spanish gentleman in bearing, manners, and dress. He spoke no English but his son had perfect English, speaking much better than I in my own language. He told me he had been three years at Cambridge and translated all my remarks to his father. As we came to the doorway of our inflatable cinema, at least 20 men immediately stood up; never mind if they blocked the view of those behind, standing stiff and straight until we turned away. They were his tractor drivers in matching green overalls. I was told that this man was second only in their lives to General Franco, having almost the power of life and death over them.

I was under constant pressure from the leader of the Spanish team to telephone England to ask for an important manager from Ford to come out to one of our demonstrations. I felt they wanted a top manager to come out to emphasise how important Motor Iberica was in the eyes of Ford and how important the Spanish market was. I did telephone our office and was left in no doubt that no one was ever likely to come out to Spain for a visit and anyway they made their own tractors which were of no interest to Ford. Before passing this information to my Spanish friend I gave a few hours' thought as how to present the problem. It seemed to me that if I admitted failure in persuading a VIP to visit them my own standing might have been in jeopardy, so I decided to tell the Spanish team leader that it was impossible for any one to visit from England because the main tractor selling season was in full swing and there was pressure on everyone to try and ensure the Ford market penetration was superior to that of Massey Ferguson in the UK. I felt they would understand this. I then went on to say that anyway I was a senior member of the sales promotion department and would devote more time to public relations now I had my team running smoothly. I think they probably recognised this as the rubbish it was but since there seemed no other way out we all played along with it and built up my image within the company to something that it never was.

I made myself much more available to them and did considerably less work on the field, only demonstrating the radio tractor, which in any case helped with the image I was trying to create. Also my appearance was spruced up as I kept my clothing smarter (and particularly the neat blazer with the Fordson wheatsheaf logo on the top pocket). If our team needed more assistance on the field I had only to ask and a small army of Spanish workers would appear as if by magic. At every location I was now into long luncheons and interviews on the radio and was also able to meet the best bullfighters in Spain because the timing of our demonstrations was made to coincide with the local fiesta, when, of course, much time was given over to important bullfights. One of the Spanish team was always present as interpreter and so I was able to learn a great deal about Spain, which probably accounts for the fascination I still have for that wonderful country.

It soon became evident that our Spanish tractor demonstrators were very good indeed especially in coping with the rock-hard ground we found ourselves on. We were told it had not rained on this site since February and it was now August and very hot so you can imagine how hard the ground was; nevertheless they were putting up a very creditable show. There was also a part for the local farmers to play: they could enter their own Fordson or Ebro tractor in a driving competition, the first prize being a trip to the Smithfield Show in the following December. This was a prospect almost beyond comprehension for most tractor drivers, who were only being paid about £3 5*s* 0*d* a week at this time and anyway were not usually allowed out of Spain, passports being almost impossible to obtain under the Franco regime.

The driving competition was quite something to watch. It was conducted in two parts. The first one was four tractors lined up on a starting line each with a two-furrow plough on the back. They stood there with

the throttle wide open and clutch held down, in what they hoped was the correct gear, facing a run about 200 yards of clear ground. At the starting signal the foot came off the clutch and the plough dropped into the ground. Talk about a Grand Prix start! The first one to the finishing line was the winner and could pass on to the next part of the contest. If the driver had selected the right gear the tractor speed would be such that the soil from the mouldboards of the plough might be thrown as much as 6 ft, certainly not a very scientific contest and undoubtedly dangerous but highly entertaining and much appreciated by the spectators. The only disaster was that at one event a tired old tractor cried enough and put a big end through the side of the engine, spraying oil over a few of the spectators. This first part often lasted quite some time since as many as 50 tractors had to be split into fours each to receive its chance in the ploughing demonstration (Grand Prix).

The second part was an obstacle course, zigzag in layout, at the far end of which the driver had to dismount and drink a bottle of Coca-Cola (most of the contents he poured over his face with his mouth open). Halfway down the return run was a straw bale which the competitor had to do a complete circle around – usually, as it worked out, in top gear and on one rear wheel. Desperately dangerous, but we who were not driving enjoyed it very much. I will never know how it was that no one turned over, especially considering the plough was still fitted to the tractor but now carried on the linkage and swinging around as the tractor violently swerved down the course. . . . All good Spanish fun.

This demonstration tour was the Spanish farmers' introduction to the Fordson Dexta. It had not been marketed in Spain before, so our equipment was concentrated on

that model. In addition to the radio-controlled Dexta we had the cinema showing Dexta films in Spanish, the fingertip Dexta (a full-size Dexta atop the extended index finger of a fibre-glass hand beneath) and a cutaway Dexta (showing internal parts). With the Spanish equipment, balloon race, driving competition, exhibitions and demonstrations of Spanish machinery followed by dancing in the evening it was quite a show. Thousands of visitors came, encouraged because entry was free. Always representatives of the local radio stations were present and I spent many hours talking through an interpreter to them; in fact I became able to converse a little in Spanish but never did so in case I said anything wrong, simply relying on the interpreter to put it over in Spanish correctly; but by now I could tell if he missed anything out.

My new life became very tiring. I was giving my Spanish friends the support they desired, but long meals where most of the conversation was in Spanish, hot days, and of course the constant travelling took their toll. But on this kind of work one is not supposed to weaken, and indeed I did not; I like to think I gave good value for the money and trust Ford were putting into me. Betty was telling me that our income was soaring but we were both unhappy about the income tax my exhausting labours were allowing a mean Chancellor to collect from my salary.

There were compensations of course. I always had one of the best seats at the bullfight and was able to learn all the finer points of this undoubtedly cruel sport, consoling myself with the fancy that if the bulls and indeed the cows that we eat were cleverer than the human race then they would be killing and eating us. I met many important people, politicians, provincial governors, police chiefs and those that most people forgot, the humble tractor driver, he who survived the heat and in some places the awful cold high up on the sierra, he who survived these conditions for a pittance of pay and who also had the skill to use his tractor in a much more efficient way than many of his so-called betters who paid his wages.

This tour I enjoyed more than the others not just because my standing was higher but because of the Spanish staff, who left other casual employment to come and work for Motor Iberica for the two or so weeks we were in town although the pay was only about 12*s* for a 12-hour day. Spain was the only country, including my own, where honesty reigned supreme.

We saw many throwbacks to the Spanish Civil War: cripples, crutches, men without legs propelling themselves about on old baby-carriage wheels by pushing on the ground with leather-gloved hands. I once sat by the mayor of a town in the west of Spain who, speaking good English, told me that one of the British brigades fighting in the Civil War had rounded up the local resistance, including his own brother, into the bullring and machine-gunned them all. A demonstration near Toledo took me to the Alcázar where I was shown pictures of a magnificent building almost blown to pieces. By now rebuilt to its former splendour, it had been the site of a terrible siege in the war. I was shown the hospital beds in the cellars of this building, the Harley Davidson motorcycle, less rear tyre, which had been used to drive a small cornmill with a belt driven from its rear wheel. Worst of all was the office where the commander of the besieged Nationalists took a message from the Republican attackers to the effect that they had his family as hostages and particularly his 13-year-old son who they were threatening to kill unless the defenders gave up and surrendered the city. The emergency telephone that was the only link was called into play and the commander of these defenders was able to tell his son to die like a Spanish gentleman. A tragic tale and one the Spanish people will never forget.

We had lighter moments of course, such as doing a bullfight in the famous bullring in Cordoba with our radio tractor. The ring was part-owned by the local dealer, who was keen to show this magic machine to the local population. Obviously we were pleased to be of service to him but unfortunately we were not allowed to try the tractor in the ring before the bullfight afternoon. Inspecting the ring beforehand showed no specific problem so I decided it would be OK and prayed no one was using any electrical or radio equipment that might cause our tractor to run amok. This was a big bullfight with some of the most important bullfighters engaged to appear. The hour arrived and I took the tractor into this cauldron of a ring with 10,000 people looking on and it stopped. Again and yet again it stopped. I am now deep in trouble. Much Spanish language was flowing, good or bad I knew not, but was only too aware of the slight rustle of impatience from the crowd, who only wanted to see the bullfighters. The problem was quite simple. I knew what it was: the stop control on the tractor was being operated by some outside means and so was not under my control, but this was our emergency stop to be operated if the tractor ran away on its own accord. The operating controls working on another wavelength seemed OK, so in desperation I disconnected the emergency stop and told my companion who was with me if I signalled to him immediately to run to the tractor and stop it manually.

The tractor now worked perfectly so we carried on with our original plan – with my companion using a Fordson flag as a cape we carried through a few passes and in fact got quite a good reception from the crowd, even a few cries of *olé* as the tractor charged the flag. Eventually we were told that was enough and to applause from the crowd we left the ring. On the way out we had to go through a tunnel where the bullfighters and their quadrille were already assembled in preparation for the pre-fight parade. I carefully guided the tractor between them down the narrow passage left for us, came out into the open behind the bullring and only then realised I could not have stopped the tractor if any problem had developed in the tunnel. Just think, I might still be in a Spanish prison instead of writing this story if one of their beloved bullfighters had been injured.

On one occasion we were travelling between Barcelona and Valencia on the N340, what they regarded as a trunk road. We thought it equated to a cart road at home, but in the early 1960s Spain had not very many resources to waste on roads that did not have much traffic. Often we had to slow almost to walking pace to ensure that nothing on our vehicles was broken. Today of course this main road is a very good one as well as having a motorway running alongside it, but the modern tourist can have no idea what it used to be like. I still occasionally drive the old road through the villages and still the men in working clothes sit outside the bar at dusty tables with bits of paper blowing round their feet, the women dressed all in black are there as before, knitting or sewing, always sitting with their back to the road, taking advantage of the smallest patch of shade. These little places are still Spain, whilst the holiday traffic rushes lemming-like to destinations further south, passing swiftly and all unknowing of these lovely villages as it pays for the dubious pleasure of cruising at 70 mph on concrete ribbons with bare rocky scenery and little character.

Just after midday there was a terrific thunderstorm. In many ways it was quite frightening; we could see lightning striking the ground at frequent intervals. Then came the rain. It was quite impossible to see so torrential was it and the noise on the van roof was deafening. We just had to stop. Soon the road was awash with water. After the storm had abated we carried on a short distance but then came to a river where several lorries were waiting to cross; there was no bridge, the road just ran through the (usually) dried-up bed of the river but now it was a surging torrent and no way could anyone cross. It took two hours before the water went down to a reasonable depth so our vehicles could make their way across. By this time it was obvious that we were not going to make Valencia that night. It was 8 pm, dusk was coming, and we were little more than the halfway stage of our journey. Our Spanish leader decided we must find a stopping place for the night.

We were approaching a small town called Vinaroz where the main road still passed through the town. (It later had a bypass, then a better bypass and now has a motorway to bypass the bypasses. They call it progress.) We parked our rather long convoy alongside the beach, just separated from the sand by a low wall. The road is still there but now it is festered up with no parking signs and has a paved walkway between it and the sand, we were told we would eat here whilst one of the Spanish team found accommodation for the night. Stepping over the low wall we sat down at a beach café, just poles with a reed roof for shade. It was in darkness but an almost full moon shone on the sea, a really lovely setting, the reflected light from the sea lighting up hungry and tired faces. Soon people began to arrive, bottles of beer appeared as if by magic, lights were turned on, a small car arrived with trays of fish direct from the harbour at the end of the beach. We had an excellent meal laced with all the wonderful flavours Spanish cooks can produce, well washed down with local wine, a meal that stands out in my memory.

The next morning we resumed our journey, having stayed in a number of private houses for the night and received a good breakfast of *churros* and rolls with strong coffee. Even today we still visit Vinaroz but alas it will never be the same again, as it is now a big holiday town well dotted with flats and villas. Leading our four trucks and following one of the Spanish Ebro vehicles there was one of the typical Spanish donkey carts trotting along the roadside on the strip allocated to them as they trundled off to work in the fields, vineyards or olive groves. In this area it was olive groves. We had been travelling about three hours and maybe the driver of the Ebro was dozing; anyway he hit the donkey cart on the left rear corner, the side of the cart was removed and the whole lot, cart, donkey, driver and most of the broken side departed like a NASA rocket into the olive trees and just below the road level, leaving the lunch basket the driver was obviously taking to the field workers spread over the road – bread, salad, garlic sausage and a small barrel of wine. This barrel amused me. It rolled across the road gently wobbling as the contents swilled around inside it, rolling straight under a six-wheel lorry coming the other way down the road. I waited for the crunch but the barrel with its contents intact came out the other side quite undamaged. It flashed through my mind that it must be communion wine. . . . The police arrived, then a lorry to collect the debris with its shocked but unhurt driver. The donkey was man-handled into the lorry by the simple means of four men, one to each leg, lifting it aboard, where it was tied to the front board before all was driven away. Thoughtfully we left the scene, still on our journey to Valencia.

Arriving in the city and as usual doing our traffic-blocking act we came to a wide bridge over a dried-up river bed. To my surprise we drove down a ramp and parked on the gravel below. In view of our experience of the thunderstorm the previous day, I enquired if the trucks were safe there, and received the assurance that they were and that this was our demonstration site for the week. Should it have worried me? Well, not really . . . but it did. Our hotel was not far away so off we went for yet another lunch, and afterwards set out the site under the eyes of many people who occupied the bridge. It worried me that in the big city with more than its share of poor people we would have a security problem – and anyway in big cities aren't there always criminals looking to make some spare cash? We stayed a full week and never lost a thing. Some of the workers brought in to help even came to ask if they could take away stuff we had thrown out. Though Spain was the poorest country we had visited it must have been the most honest. It has changed now, but of course Franco has also gone, his stringent laws have been changed, the do-gooders have arrived.

One of the characters of each demonstration was a fellow who arrived the day before the event in an Ebro truck propelled, as they all were, by a 4D Fordson Major type engine; but this truck had a red lightning flash painted on the side of its canvas tilt with the words 'DANGER EXPLOSIVES'. It caught my interest the first time I saw it, taking me back to those early days when a friend and I had made explosives in the war to provide us with a few extra rations. I checked out this truck as soon as possible and being told that the owner was one of the leading Spanish fireworks experts soon whetted my interest. He later told me he had been invited to provide a display in England at the last coronation. I don't know what reaction he received on that occasion but certainly now he would have been 'run out of town'.

However, he was quite a magical character to me, sitting on the open back of his truck with a cask of powder, about the size of a five-gallon oil drum, nonchalantly pouring powder into paper or cardboard tubes, preparing display pieces for the evening performance; even the words 'Ford' and 'Ebro' were being fashioned in coloured fire. What worried me a little was the cigarette he was smoking at the time he was showing me how to do this and telling me both his father and grandfather had been killed by firework displays. His speciality was what he called the waterfall. This was a line of rockets some hundreds strong with differing length fuses so that as he ran along the array with his taper lighting them they would all go up together. The framework they were

suspended on had twice fallen over (thus the tragedies in his family). He was convinced that his now modified framework would not fall again as it had on the previous occasions when the last of the rockets with almost instant firing fuses had fallen over and 'shot' both his father and grandfather.

I had my doubts. At a later demonstration in a town called Badajoz near the Portuguese border it was decided to hold a grand dinner in a fine hotel bordering the town square and afterwards culminating in a firework display in the square. After dinner we all trooped out of the hotel to enjoy the display which started with the most ear-shattering bangs shaking the ground and giving a bright flash and crack in the sky. All went well until it was time for the waterfall. No, it did not fall over, but what happened should have been entirely predictable. A night with no wind . . . all the rockets went up OK but of course the debris had to come down, straight down. As everyone ran for cover the rocket sticks were falling on people and car roofs making a noise like machine gun fire as they hit the roofs. It was almost as exciting as the waterfall itself.

Eighteen months later we were on holiday in Denia at fiesta time and, yes, there was a firework display and, yes, it was our friend preparing it, sitting on his truck on the harbour wall. We met as brothers but at the display that night I made sure our party stood under a palm tree to watch the waterfall.

As our last event drew near we assembled at Salou south of Barcelona, where many of our Motor Iberica friends from the factory were present. Finally we were entertained to a grand dinner at a small fishing village called Cambrils, where a very good restaurant gave us a wonderful *paella* right on the waterside.

The next day as we were preparing our equipment for the journey north and home the sales manager asked me to learn Spanish and return the next year as their sales promotion manager. Quite a compliment, I thought. But of course there was to be no return; my opinion then, as now, was that if you cannot earn a living at home it is no good trying in another country where there are inevitably handicaps you never thought about. As we loaded our truck the thought came to me that this was my 64th farming fair – and unknown to me it was to be my last.

As we crossed the Spanish border and headed into France there was time to recall all the enjoyment of the trip we took to Gibraltar one weekend when there was no demonstration; how we had hired a taxi, clubbing together to meet the bill for a hundred mile journey and payment of the driver's overnight stay so he could drive us back again. (At that time Spanish people were not allowed to leave Spain to enter Gibraltar, so he had to stay on the Spanish side of the border.) We were taken by a very old Citroën taxi over the border, there being very few that were allowed to make the journey. This one was positively the most unroadworthy car I have ever ridden in;

the driver had to continuously correct the steering, use the hand brake to stop it and had a piece of wood to jam under the gear lever to hold it in top gear. Luckily the journey was only about one and a half miles.

Having stayed in Gibraltar for just less than a day we decided to telephone our driver and tell him we were staying another day, to enable us to take a day trip we had seen advertised to Tangier. None of us had ever visited the African continent and here was our chance to go.

It turned out to be quite an adventure. We were met off the plane, after a 15-minute flight, by a great crowd of guides and taxi-drivers. We only had a day there so it was decided we should use a guide . . . but which one? We had heard tales of the guides and how they would rip one off. We selected one who had a large enough car to hold our party and I must say he was very good value, taking us around town, out into the desert to ride on a camel, to the most northerly part of the country, the so-called horn of Africa, back to town for lunch, a visit to the *souk*, or market, where we experienced the local pressure-type selling – all very pleasantly done and wonderful to see. He also took us to see a snake charmer and the brave amongst us had photographs taken with snakes adorning various parts of our bodies. (No, I am not brave.) Eventually we returned to the airport for our return journey, having paid our taxi man £25 amongst four of us, so it turned out to be very good value.

At the airport we were told it was too windy for our aircraft to take off and we were to be put in a hotel for the night. . . . Disaster. Our taxi-driver was expecting his bill to be paid that night and we were stranded in Africa. We telephoned and with great difficulty managed to arrange another night's stay. By now we had collected a party of 10 persons, Americans and Canadians all awaiting the next day's return flight to Gibraltar. Having charged us £1.10*s* for the day trip British European Airways must by now have been on to a pretty well-used bad wicket; now they were paying for a hotel and dinner.

After dinner we decided to look around town and while doing so decided a beer would be a good investment. A bar was found and in we went. Most unusual – the first thing we found was that the two bartenders, possibly girls, with quite short skirts, were sitting on a chair having a snogging session. We eventually got a beer each and sat at a table one chair short of ten. I found a chair at a neighbouring table and with a gentlemanlike request asked if it was free. Possibly removing it before the people had time to answer, I took it. Immediately one of the three customers sitting at the table pulled out a knife. 'Put it back,' he said. I looked at him and in a film would have kicked the table in his face. I am not that brave (or is it foolish?) but mildly replaced the chair and found another. The three sat glowering at us. How good it was to be a party of ten!

The next day we returned to the airport but it was still too windy for the plane, so we were transferred to the sea ferry and had a good trip back.

There were so many memories of Spain. On the central plain (where the song says it mainly rains) there were some very large farms indeed but often nearer to the coast the farms were much smaller family-run holdings, usually worked by hand in those days, so we would see men scything fields of corn or cutting it with a harvester. Here was rural life as it must have been in England a century before, but here they had the weather to help.

I noticed that when the corn was cut it was not stacked as we would have done but moved, with a mule and cart, to a large round circle of hard-packed soil, where it was unloaded; and when the quantity of corn was sufficient, a mule and a sledge with a chair securely strapped to it were driven round and round until all the grain was rubbed out of the straw. Remember, I said the weather was on their side; being so dry the grain came out well from the husks and the straw would often be broken up as well. It seemed to be Grandmother's job to ride the sledge. Always dressed in black, these old ladies would sit, going round and round most of the day. Eventually perhaps after half an hour's rubbing they would take forks and by throwing the rubbed mixture into the air would extract the grain, the wind blowing away the chaff. Any straw that did not blow away would be lifted and shaken again, to be sure no grain was left in it, and dumped at the side of the circle. By this time the next load would be ready for rubbing and so off went Grandma again, the remaining straw being put into an upright box and by a system of ratchets compressed until quite presentable bales were formed.

The other things I loved to see in Spain were the waterwheels used to lift water for irrigation. It fascinated me to watch the mule walking round and round driving the wheel, with earthenware pots strapped to a rope that conveyed them to the well and dipped them into the water; and as the pots turned over the top of the wheel the water fell into a trough that carried it to irrigation channels. The waterwheels were very numerous at that time, often only a few yards apart in some places; but now, alas, gone forever.

My Spain is no more. I often go there but only memories recall the sights seen so long ago. The main thing that made that Spain work so well was the sunshine and that is still there of course. They call it progress: now the water is lifted by electricity and the grain threshed by combine. Progress has to be. The people are much more prosperous; not so sunburned as those who threshed the grain in the old days, but still hard working. Franco is gone, but as a friend who had lived in Spain said, 'When Franco was here I could leave my car in the street all night, even with my briefcase on the rear seat and unlocked and know it would never be touched. Now even with the car locked it is likely someone will break the windows and steal the car, or take the radio. Democracy as every thing else certainly has its price.'

6

I NOW started what was to me almost a new life, spending much more time at my desk in Ilford. There was a ploughing match award scheme to help administer, new dealer openings and established dealers who wanted to push for a little more publicity. Often I might be involved in those events and be away from home for a time, usually only for a few days, taking the radio tractor along. And so our life in Heybridge took a better and much more settled and enjoyable turn, many of our friends and relations from Derbyshire being able to visit us. We had outrageous parties. We had settled well in Heybridge and many of our neighbours joined with us. I well remember the milkman calling one morning about 5 am and saying 'You are up early. Are you going to the seaside?' No, was the answer, we have not gone to bed yet.

During the whole of this book I have not mentioned Mr Woggins, our black spaniel. He also lived in Heybridge for quite some time but alas whilst I was in Spain Betty had to write the sad news that he had had to be put down. His back legs had failed. We think he was 15 years old. I sat reading the letter in my hotel room, thinking of all the times he and I had been together, at ploughing matches or taking sugar beet to the factory, all the hours we had shared together on David Brown, our old tractor. It was a sad day for me but I guess even sadder for Betty who had lost part of the family we had shared in our earlier life. I knew she would soon recover but perhaps never forget as she settled in new surroundings with Nicholas.

About this time Ford was changing; a number of senior figures were being moved across the Atlantic to America, becoming part of Ford International. There was much speculation about who might be next, and the wonderful team spirit I had instinctively felt when first joining the company seemed to me to be wasting away. There was beginning to be evidence of strict financial control imposed from America. Rightly or wrongly I felt they were quite mad; the bean counters were just beginning to flex their muscles. Eventually I would become so incensed with what to me seemed insane management that I would begin to look around for a better future; but at least for the time being the work I was doing was very enjoyable, especially so now it involved working with so many Ilford colleagues.

In the summer of 1963 began my involvement with the main agricultural shows. It had always been Bill Baker who was the prime mover and planner of the show stands at the Royal and Smithfield shows as well as others. Now I found myself working as a kind of assistant to him. It was all very interesting especially so because, although we were deadly enemies of the David Brown, IH or Nuffield staff, in the evenings we could enjoy a party with our competitors, as anyone would who either stayed at the same hotels or used the same pub. The days of reasonable expense accounts and no Breathalyser laws allowed us a great deal of freedom, just a little abuse of motor cars and some tyre wear as the road was burned up when the pubs closed. We often benefited from the local pub's dubious ways of staying open until the early hours of the morning when a good party was in full swing and the till was ringing its heart out.

Our show circuit comprised the Balmoral Show in Belfast, Royal Highland, Bath and West Show, Royal

Show, the Royal Welsh Show, then at the end of the year the Smithfield Show in Earls Court. In the meantime I was looking after the British national ploughing match, and four smaller shows, in addition to whatever turned up and might be described as a sales promotion function. I still used the demonstration van a great deal to carry show materials about. It was often quite a frustrating experience being subject to Bill's overall control but I have little doubt it was good for me and stopped my natural enthusiasm from running away with me.

Our Tracteuropa trucks were now only used in the UK and even then mostly for haulage purposes. You may wonder why the continental demonstration circuit died so suddenly. Well, things change in companies and so it happened within Ford; Ford International were beginning to control more and more of the overseas markets and in their American way were applying pressure to each Ford company in the countries they controlled, to produce ever-increasing profits. If one company did particularly well the others that might fall short were asked why and so each company became its neighbouring Ford company's biggest competitor. Thus the collaboration enjoyed by all when Dagenham was mainly in charge disappeared and such things as a Tracteuropa tour to help out a neighbouring country were counter-productive to Ford of Britain's well-being. Things were changing fast.

We still enjoyed considerable freedom in our show work. About this time a chap named Tom Jamieson joined the company. He and I hit it off at once. He was a Highlander from the Inverness area and a very fun-loving character. Often he was on show duty with me, especially when Bill Baker was seconded to America to help in the announcement of the new Ford tractor (alas, the name Fordson was about to be lost to us). Bill spent almost a year in America and John Prentice became my supervisor again although he now spent most of the year in the office whilst I became his field man, so the previous year's experience of working with Bill on the shows was now paying off.

Around the shows all summer it was inevitable we should have some adventures. Our stop in Edinburgh for the Royal Highland Show is remembered with some pleasure, not only for the work we did, which I tried to do as efficiently as possible whilst at the same time, hopefully, retaining a relaxed spirit within our team. We were booked into a rather classy hotel in Charlotte Square, The Roxburgh. It was nice to live well once again, but we found the dining room, although excellent, too expensive for our daily allowance and although we could no doubt have worked the bean counters over and recovered extra monies we decided to find a rather cheaper source of nourishment that might just allow us to enjoy one of the many good pubs in the locality without actually breaking our personal bank accounts. We eventually ended up as a party of ten eating at the Greek restaurant around the corner from our hotel. We soon had them trained. As we walked through the door our waiter would produce ten bottles of chianti on the reserved table set for the ten of us. We ate and drank well in that restaurant, enjoying the music, usually Scottish not Greek, and at reasonable cost. It became a matter of pride that on the way back to the hotel someone climbed the no-waiting sign at the end of the street and bent the round top over so it could not be read. Tom was good at this and we found out later very good indeed with the bagpipes, his family having a tradition of piping.

On another visit to the show we spent a fair amount of time in South Queensferry whilst still living in the

Roxburgh Hotel so had to drive back to Edinburgh in the evening. After dinner one night (in fact it was probably early morning) we were driving through the rather wooded countryside where we persuaded Tom to play his pipes. Up to then we had not known he was able to produce music on the things, although he always carried them in the boot of the car. I guess that night he had drunk enough whisky to believe the fumes would keep the air bag supple enough to give a good tune. In any case he produced them and after a few animal-like noises as they were warmed up he serenaded us in the quiet of the Scottish countryside. Indeed he was very good and without exception we enjoyed his serenade. Eventually running out of steam he stopped. The woods were quiet no longer – crows, pigeons and birds of all sorts were volubly protesting at this unwarranted disturbance of their slumbers. We departed for our hotel.

During the installation of the Royal Show at Stoneleigh several friends came to visit and were persuaded to stay for dinner. We were staying at Brandon Hall hotel, where we enjoyed a long alcoholic dinner and afterwards in the bar were told of a ghost who sometimes could be seen in the old coach house out in the wood at the side of the hotel. Now it was foolish to tell a well-oiled tractor demonstration team about a ghost and not expect them to do something about it, so we did. Tom produced his pipes and we all trailed through the nettles and docks into the coach house, led by this wailing demon of a piper. Forming a circle in the coach house lit by various torches we were treated to another of Tom's concerts, but no ghost appeared. However when the pipes ceased to play we could hear, in the distance, a deep woof-woof. We were not to know the manager had a St Bernard dog who did not like bagpipes. There were some residents in the hotel who had the full force of the dog's voice, so a mixture of pipes and dog at 1 am was not well received. The next morning we had a complaint from the manager about our behaviour and so to help smooth matters over we gave him a few free tickets for the show.

This was another disaster. The first day the show opened he arrived on our stand during the afternoon in a very happy state, obviously having visited some of the other stands which had personnel staying in his hotel and who were only too pleased to provide refreshment for special visitors. There was a certain amount of commotion culminating in one of our managers asking 'Who the hell is that?' It was time for action, so one of our company was detailed to escort him back to his hotel. We never saw him again; the next year there was a new manager.

In those days Brandon Hall had a putting green which we used, so the next year we decided to behave ourselves. It was a good hotel and we did not want to blot our copybook to the extent that they might be tempted to refuse a future year's booking, so it was decided to have a grand putting competition. After dinner we arranged about ten vehicles around the green for floodlighting and had a very successful evening. As we retired to the bar to drink the health of the winner, a sporty Jaguar car with two slightly upturned exhaust pipes was parked facing the green. I just could not resist rolling two golf balls down these tempting targets. I never heard any more about them and so am left to wonder if the Jag had a funny exhaust note or if they were perhaps propelled out of the pipes at speed down the motorway. Another of our exhaust tricks was to have a small bottle of sump oil to pour down our competitors' tractor exhaust pipes as the tractors stood on the show site all lovely and polished. If a customer required the tractor started up during the show

we could always tell by the amount of oil smoke around, not to mention the black spots that covered the tractor and (who knows?) even the customer. But then, what sensible customer would have been interested in any other tractor than a Fordson?

At the minor shows there were always some tractors on loan to other manufacturers so they could display their equipment in working form. Often at the big shows we might have as many as 60 tractors on loan but at the small shows perhaps 12 might be a reasonable number. The reason for this was not charitable and a gift to the other manufacturer but was a conscious attempt by Ford to flood the showground with our tractors to emphasise how versatile they were; and remember at a main show we would have 25 or 30 tractors on our own stand, so a good representation at these shows was assured. It was always a big job to prepare these tractors for display. They would come from the factory probably after standing in the industrial fallout from Dagenham with preservative covering the most vulnerable parts and so we had a big job to prepare them for display, especially on the small shows where I was usually working on my own, and in addition to preparing two tractors for our own stand, I had to deal with a 12 × 12 tent and all the display material and leaflets the showstand called for, as well as booking my own accommodation and arranging transport. In the meantime Sales Department would be trying to place these slightly used tractors with local dealers at a small discount at the end of the show. They would then notify me of the sale and quote a serial number for the dealer to collect or for me to arrange transport for; alternatively I might be asked by sales to dispose of the tractors to dealers in the locality and then notify sales where they had gone. There was of course a discount for the dealer to make these slightly used tractors an attractive buy.

One of these small shows, the Public Health Inspectors Conference, was held in Eastbourne, on the local rubbish tip. I spent a week preparing tractors and our showstand, then it started to rain and continued to do so for the two days of the show. The state of the working tractors had to be seen to be believed after trailing around a rubbish tip for this time. I had to dispose of them and yet felt a need to send them away in a reasonable state – after all I might need to dispose of some more in the future, or indeed meet the dealers at some future event. I did not fancy cleaning these machines myself. Apart from the hard work, which I could stand, there was the next show at Southport to consider, where I was due the day after this one closed. Eventually I had an idea. There was a car wash not too far away so tractors dripping with mud could be seen going through the car wash with the attendants shovelling mud away before a motorist came with his car. I paid them £5 to wash each tractor. It was money well spent.

It is always better to be lucky than rich (so they say) and I felt lucky when the Groundsman's Conference decided to have their show in the Isle of Man and it was Manx Grand Prix week. So again I was off to the island, this time managing to work a trip for Betty and Nicholas as well. Our showstand was in the Villa Marina gardens on a nice lawn and I had only three tractors. The demonstration van was booked across on the boat from Liverpool so we arrived in good time . . . only to find the van would not fit into the side opening of the boat. A quick bit of work removed the luggage rack off the top and deflating the tyres enabled the van to scrape inside, with four heavy dockers sitting on the back to lower it further. We had a wonderful week in Douglas, where the Palace Hotel was our home. We resisted the opportunity to gamble in the casino but did drive the demonstration van around the 37¾ mile T.T. course. I realised then that the van had been round this course, the Nurburgring in Germany, Silverstone and the French Grand Prix Course at Reims – quite an achievement for a van intended to help demonstrate tractors.

One of my favourite shows was the Balmoral Show in Belfast. I could not really say why but I think because the Irish are so laid-back. Also, I had been over as a demonstrator and had enjoyed their company very much. Here I was again meeting with those very people. No doubt another attraction also was the nearby bar where one could hear the most outrageous tales after a couple or so pints of the 'black stuff' (Guinness). I stayed in The Royal Avenue Hotel and parked my van in a street at the side, never suspecting that a few years later to have a van with 'Ford of Dagenham' on the side would have almost been suicide. But even at that time, after I had eaten in a restaurant with a friend working for the main dealer and offered to run him home, he refused on the grounds that if I brought that van into his district I would never get out again. That made me a bit careful where I went afterwards.

As the lives of the Major and Dexta models were reaching the end, an important dealer meeting, to be addressed by Mr Batty, was held in London. There the dealers, who were already unhappy with the supply of tractors for the home market, were told that they would face further shortages due to the production moving from Dagenham to the new purpose-built factory at Basildon, no mention being made of any changes in tractor models. This meeting did give me the opportunity to say, as a member of the organising team, that I had stayed in the Dorchester in Park Lane and eaten in the excellent restaurant there, being present at the banquet after the meeting and laughing along with the Ford dealers at the cabaret, featuring Nicholas Parsons and Cardew the Cad making fun of our tractors, much play being made of the double-acting ram valves and other features being mentioned in a context no one in the Ford organisation had previously thought about.

Another adventure was the Irish national ploughing match. For some reason they decided that the radio tractor should be present and although Ireland and Ford Cork seemed to be separate companies it appeared that the UK had a connection still and so, although it was no longer the policy to assist foreign countries to sell tractors it was OK to assist Ireland. So whilst I looked after the British ploughing match in the Wirral in Cheshire the radio tractor went on a normal tractor shipment to Cork and later was moved to the ploughing-match site not far from Dublin. Flying from Liverpool to Dublin I was reunited with the radio tractor and after the transmitter battery was charged the tractor tested out OK, so off we went to our hotel, the Ford Cork service man and I. The next morning it was snowing very heavily. It took us three hours to reach the ploughing site, which itself was well covered in snow, so there would obviously be no ploughing that day. The organisers were forced to postpone the event for a week, the dealer 15 miles away agreed to store the tractors and the radio tractor was taken on his lorry. We were left on site with 20 tractors to move. They were not my responsibility – I only had the radio tractor to look after – but being a Ford factory man there was, I felt, a loyalty to the system which says we all help one another, so I volunteered to drive a tractor on the 15-mile journey.

Whilst the organisers were deciding what to do about the weather the dealer invited several of us into his caravan for a drink. It could only happen in Ireland. There was a galvanised pail on the gas ring nearly full of soup, bubbling away gently, bowls were handed round and a large ladle of soup dished out to each of us. After the cold outside the soup was wonderful and there was also good rough country bread. The dealer then opened the caravan wardrobe and we found it was half full of gin bottles so we each had a good measure of gin poured into our soup, the allowance seeming to be about one bottle of gin to eight bowls of soup. Never has soup been so warming in a snow storm!

Four hours later, another bowl of soup accompanied with Irish stories and jokes and we were ready to drive the tractors away, at least some of us were, I think. It was almost dark and snowing even harder, even then the question went through my head, 'Do I really want to do this?' But in Ireland one either joins in and enjoys the experience or remains dammed forever as a stuck-up Englishman, so I drove a Major, in the dark, in a snowstorm, standing astride the seat (so my face was above the torrent of snow flakes diverted by the bonnet up and into the face if one sat on the seat), one hand held trying to shield my eyes from the snow, the other steering the Major, lights on and reflecting back into my eyes, driving almost blind on a road I did not know and trying not to lose the tail light in front. The cold was desperate; at the end of the journey I and several others had difficulty in walking and I assure you it was not the gin. A week later I was back in sunshine with a trailer-load of passengers and instructing the Minister of Agriculture on the finer points of driving the tractor from a seat in the trailer . . . and still there was gin and soup for us in the caravan.

The International Horse Show at White City was quite a different event. Our job here was to provide tractors and trailers to move jumps around in and out of the arena. We were needed to prepare the event from 8 am and the last event finishing at around 10 pm meant we were not away from the place before about 11. We had a good hotel to stay in but even so spent very little time in it. The mainstay of the course builders' staff was the Junior Leaders Regiment. The course builder was a gentleman by the name of Talbot-Ponsonby, a wonderful man who certainly knew his competitors and horses, an ex-army officer and a strict disciplinarian. The Junior Leaders were given hell, constantly being corrected and told to do things at the double, all part of their training I am sure. Even we addressed him as Sir but the next year I made definite progress, calling him Jack whilst I was no longer 'you', my name was Arthur. It was very interesting to be at close quarters to the jumpers and observe their differing characters, the non-stop preparation of the horses and the devotion of the grooms, not to mention the almost all-night parties in the back of horseboxes.

The motive for the expense of providing this service to the event was of course publicity; the event was often televised and Ford wanted a slice of this publicity by providing tractors which could be seen on television moving jumps around between events. And for this purpose what better to use than the radio-controlled tractor? Television producers always want to find unusual things to put on the screen and what was more unusual in those days than our tractor? We only expected to be allowed to use the tractor on one or two occasions because Jack Talbot-Ponsonby was a real keen in-and-out-quickly-with-those-jumps organiser and our radio tractor was not the quickest mover but he agreed I could use the tractor one evening about 10 pm whilst the event was being televised. He told me that the course he would build should give eight clear rounds to take part in the first jump-off; the final jump-off he expected would have three competitors left.

As I said before, Jack knew his horses; his knowledge was almost unbelievable – he could even tell most of the names of those he expected to go clear. This particular evening he told me which jump he wanted the tractor driven to so that the Junior Leaders could load the jump on the trailer for removal from the arena. As the last horse entered the arena I checked the tractor and started the engine. The tractor refused to steer to the left – one of the tuneable reeds that controlled the left steering must have gone off tune, maybe the evening air affected it. Quickly I undid the two screws holding the radio in its case, and with the reed back on tune, Jack shouted, 'Now!' No time to screw up the radio, just drop it into the case and go. All worked well. I steered in the direction of the royal box, stopped the tractor and saluted the Queen by flashing the headlights and with millions of people watching drove to the jump, collected the pieces and drove out of the arena. As the tractor came out, the steering to the left failed to work, but we had about a minute and a half of free TV time. 'Good show', shouted Jack. If only he knew how near we were to disaster. . . . These are the things that breed ulcers. Although I have schooled myself to keep calm, it is a terrible feeling to know how close we were to disaster.

7

OUR new factory in Basildon was almost ready to produce the new range of tractors, and our desks in Ilford were packed. One Friday the removal men came and Monday morning we reported to work in our new office in Basildon to find everything in place, a good move by all. The new tractor range started to come off the production line and a few weeks later our strong, some would say dictatorial, manager Mr Batty departed for pastures new. A whole management reshuffle now took place, some people departing for Brussels where a new Ford European operation was just gaining strength. Now the new worldwide Ford tractor was in production, Select O Speed gearboxes from Detroit, transmissions from Antwerp, engines and hydraulic assemblies made in Basildon, were all meeting for assembly into tractors to be shipped all over the world. To say it was a shambles would have been an understatement. A new factory with new machines and many new staff did not help. Many, probably most, tractors came off the line incomplete or defective. All this was remedied in due course by a rework programme when staff were asked to volunteer to work in the evenings to fit missing parts etc. to tractors so they could be marketed. I enjoyed this: it was back to my roots as a mechanic.

The main problem, and one never recovered from, was the marketing of the Select O Speed transmission. This problem was based on the complete inability of management (in my opinion) to grasp the basics of the unit. They were enthusiastically plugging its advantages whilst not having the service back-up to remedy any mechanical problems that came to light and lacking the operating knowledge that we demonstrators had. I feel the Americans left us totally on our own and gave little or no help; for instance, we soon learned that to change gear smoothly it was necessary to drop the engine revs when moving the gear lever but we could not persuade marketing that this needed the foot throttle they were marketing as an option to be fitted as standard. Thus the gearbox got the nickname of 'Jerk O Matic', which was totally uncalled for. Also, in spite of sterling efforts by Boreham to train dealers in the repair and service of this new type of transmission, the dealers' awareness of the service procedures was sadly behind the best customer requirements. It is easy to look back and say *if only*. . . . I believe there were enough problems with the new factory, new staff and new tractor not to have introduced this revolutionary transmission until perhaps a year later, when I am sure it would have been successful; but, as ever, Sales wanted to maximise publicity at the launch of the new machines. With hindsight I believe they were wrong.

Other problems raised their heads but again they were problems that need not have occurred. The Ford 5000 (not, you note, a Fordson) had some engine problems but the old Major was a low-revving, rather slogging type of tractor, whereas the new tractor had an engine that needed to be turning over almost at maximum revolutions to get maximum power. Totally different characteristics – but the tractor was never sold on this difference and so the engine was dubbed gutless and in addition had some piston problems which perhaps were related to these differing characteristics. In short the new tractors were not given a good chance to succeed until mechanical changes made a new marketing strategy necessary and vastly improved their image. Eventually we had a country-wide demonstration scheme to help, and the tractors became the success they should have been in the first place. Of course I, like everyone else, was quite ignorant of these problems but being at the sharp end of these things, I soon came to know them and often how to get around them. But to try to tell management or even worse the bean counters to spend a bit more money or time attending to these matters was impossible.

Talking of the bean counters (finance department) reminds me of a little story that I feel sure started the eventual dissatisfaction that began to creep into my relationship with Ford and led to me looking around for another employer. I had already been through several disputes with finance usually about how much I spent in hotels. Maybe they were correct: perhaps I was an expensive man to keep on the road. (You who read this book may have already formed that opinion.) But that was never a deep-rooted problem because I knew there was need to keep expenses in check and the bean counters were only doing what they believed was in the company's interest.

What bugged me much more was the fact that these men of little knowledge – I used to say they had not got enough sense to go indoors if it rained – were always trying to tell me to do things I knew did not work. For example, when I had washed the tractors in the carwash in Eastbourne they still did not look like new ones so I allowed the dealers who took them a £50 discount. Eventually I was summoned to the bean counters' office to explain why I had given so much of the company's money away on those tractors. There were no accusations but veiled suggestions that someone in my position could have taken a cut from them myself. I ignored this because it fitted in with my low opinion of them, but when one of them got around to lecturing me on the lines of 'Look here, my boy, if you want tractors from the factory for your demonstrations you must not give away more than £25,' my response was immediate and loud. I told him my opinion of his intelligence and the fact that I personally did not want tractors at the shows; I would be happy to sit on a chair giving out leaflets, tractors only caused me pain and grief trying to look after them, but Ford wanted tractors on the shows to help sell tractors so they could pay his wages and in any case it was not my job to sell them from show sites but that of the sales department. I stormed out, slamming his door (I guess it was not the first time it had been slammed). A few days later I received a memo authorising me to allow up to £50 discount on ex-show tractors but only if £25 could not be agreed.

The next year of my employment with Ford was very interesting indeed. We now had new models to promote and they were not performing too well. It had been decided that a sales promotion drive was needed and so a new set of temporary demonstrators were recruited. (How that took me back in time!) We also had a new sales promotion manager who had good ideas but the budgets were usually exceeded and this I feel sure caused him some pain. However, a budget was raised to provide new articulated trucks to transport tractors and so when our new team was trained a series of demonstrations was laid on. I was not part of that, being involved in the shows, but eventually it was thought a good idea to have a touring display team for the new season's shows and so the Ford display team came into being for a short time. This visited the shows and gave displays in the main ring with driving competitions between two teams of demonstrators and a public address system mounted on a new van with its own commentator to whip up interest from the crowd. It was quite impressive, especially because the trucks drove into the ring each with four tractors on it and as they drove around trailing the ramps behind them each tractor reversed off. Loading them up was even more impressive. The big tractor went up first and finally the smallest. As each tractor ran up the ramp it was travelling slightly faster than the truck dragging the ramp, but as soon as the wheels gripped on the ramp there was a sudden acceleration and the tractor jumped forward. This was OK for the first tractors but the fourth one had only enough room to ledge on the back of the truck so that the driver had a nightmare situation at each performance. There were no accidents and great credit must be given to the demonstrators for that.

The radio tractor was part of this team, but I was not the driver on this occasion. Jim Taylor drove a truck with a Brimech body (that is one designed to roll back and tilt for unloading). This meant that he could drive in the ring, tilt the truck body, drive the tractor off the truck, do his demonstration, drive the tractor on to the truck again, lift the truck body, and drive out of the ring, all without leaving the cab. All this with a commentary talking about Ford technology, etc. This amused me because that radio tractor was unusual because it had been assembled by me – a case of the blind leading the blind. The story was that when the Dexta was replaced with the new Ford tractors, management decided that a radio tractor was now 'old hat' and we did not need one so I was detailed to remove all the radio equipment and get the tractor ready for sale as a normal machine, which I did. When I asked what to do with the old equipment I was told to throw it away; but being cautious, I put the old equipment in a box in our store; and sure enough a few months later someone decided to have a dealer opening and a request for the radio-controlled tractor was made.

A change of management also brought about a change of policy; I was called into the new manager's office and asked what I had done with the old radio tractor equipment. When I gave the answer I was asked if I could install it on a new tractor. For the benefit of the new manager I agreed to try but made it clear that the old tractor had many restrictions and although there was a possibility that if a new Select O Speed transmission was used the tractor could be improved, there was no way around the basic problem of the model aircraft radio that formed the basis of this tractor. Of course we had the required licence to operate it and also notified the GPO of the location where we intended to use it, but often suffered from the problem of model aircraft users upsetting our tractor just as we upset their aircraft; and so as a long term project a new radio tractor was mooted using a new digital radio band allocated for this purpose. In the meantime I was to operate from one of our dealer's premises and with a member of his staff attempt to rebuild our equipment on to a new Ford 3000 tractor. After giving the system considerable thought I was able to use the eight channels available to us in a better way. I had the advantage over the original makers of knowing just what was best suited for sales promotion work and so our new tractor which emerged at the end of a week's work had different features to the one that preceded it. Also it was possible to use the eight channels in differing ways for different demonstrations, as they were plugged in accordingly. It was this tractor that accompanied the display team and was used by Jim Taylor as part of that team.

The end of the road for that tractor was in sight and at the end of the summer I found myself investigating the provision of a new radio tractor using this new digital radio. Although our manager had the ultimate responsibility and John Prentice the day-to-day responsibility, I found myself visiting a few suppliers whom we considered able to do the required job and making recommendations on their ability to provide a new radio having 32 channels that would form the basis of a good sales promotion tractor. When the tractor was approved and made we really had an excellent working unit, no longer eight but now 32 channels, several of which could be worked together (something we could not do with our old one), and all the gears, eight forward and two reverse, were now available. During the winter I found myself visiting dealers' open days and other functions with this toy and really enjoying it. The transmitter aerial was mounted on the van roof and so I could sit in the van and give a good demonstration of the tractor's live hydraulics and live PTO facility, coupling the Select O Speed gears with the front loader fitted to the tractor to show the versatility of all these features.

But undoubtedly the most popular and silly use was to talk about these features on the van's public address system and at some stage have the tractor charging towards the van at high speed and in a feigned panic cry 'Good God, it will not stop!', just at the critical moment putting it directly into reverse, so whilst it charged forward the wheels were, for a few yards, spinning in reverse. I never had to ask for a new van but this certainly amused the crowd. As the tractor retreated from the van I was able to put across the advantages of Select O Speed that had so obviously been demonstrated.

During the winter months a scheme was worked on to demonstrate the tractor range by laying on a 'bullfight' using tractors at the Royal Show, and previous to that during the spring to show our new radio tractor to the press and suggest ways it could be used commercially. So these two events took up quite a lot of my time during the early part of that year. For this press event we used the Hickstead horse-jumping arena and I was able to drive the tractor from the commentator's box, which was quite high enough to give good vision over the whole area. Various little demonstrations were arranged, culminating in one suggesting the tractor could be used where it was dangerous for a driver to go.

Television has recently shown a robot blowing up a car as a counter-terrorist move; well, we perhaps foresaw this type of use thirty years ago. Our demonstration was arranged so we could use the water jump in the arena, it being intended that the tractor would be used to put a charge of explosive into a hole, perhaps suggesting putting out an oil well fire. We were not wanting to arrange a 'pussycat' demonstration so it was decided we would use real explosives. (Guess who suggested that?) The arena site man was a timber merchant in his spare time and so had access to gelignite, which we deemed ideal for our purpose. The tractor had a front loader fitted and so a box marked 'Explosive' was used on the front forks of the loader. The idea was to put the box into the hole and as I did so the site man would detonate the gelignite by remote control. We were unable to use a fuse because it was impossible to get the timing correct and although we could have organised firing the charge with radio it was complicated and time did not allow, so a simple demonstration was deemed OK.

All went well with the other demonstrations, many newspapers having reporters and photographers present. The gelignite had been put in place. I drove the tractor to the hole and carefully placed the box into it. The tractor retired a few yards and our site man fired the gelignite. Unfortunately I had placed the box directly onto the explosive (it was not easy to see just where it was to the inch from a hundred yards away) and a photographer, not expecting a real explosion, had ventured close to the hole. As the gelignite exploded the box was blown to pieces and the photographer stood in a hail of broken wood with nails sticking out of some pieces, which passed him travelling at high speed. He turned out to be from *The Times*, and we got a write-up inside the next day's paper. I wonder if a bigger one would have been provided if the poor photographer had been a fatality.

During the summer the projected bullfight was worked on until at the Royal Show a bullring was erected with scaffolding and our tractors, now rehearsed and ready, were put through their paces. The radio tractor with horns as the bull, a 5000 with loader and a large red cape suspended from it to represent a matador and many other gimmicks to help show off the features that were available on a Ford tractor. Because I had been on site during most of the preparation work it was decided that the radio-controlled tractor would not be driven by me because if any one fell ill or was absent from the team the knowledge I had of the whole procedure would enable me to step in and deputise, so another member of the team, Bruce Keech, drove it for all the shows at the bullfight. I mention this in case anyone who saw this event (and thousands did) should think I was driving the tractor, when in fact it was Bruce. He did it so well I feel he should receive the credit for this.

Some problems always crop up and we had two at this event. First the crowd was controlled by attendants who also gave out Ford leaflets and were dressed as Spanish policemen. We soon had word from the Spanish

The author standing beside a radio-controlled Dexta in Seville in 1962. The Fingertip Dexta is in the background

At the Blagdon Farming Fair in 1962

Roadless-Case tractor, pre Second World War, used for launching lifeboats. Note the runner-jointed tracks

Farm Tractor Drive's first four-wheel-drive tractor, featuring side drive gearbox and 'Sige' Italian front axle

Derbyshire County Show in 1983

Chieftain publicity photo of 1984

Son Nicholas on a Super Dexta at the Farming Fair in Kilmarnock in 1961

Nicholas' children, Philip and Clare, with the same Super Dexta in 1995

The author with John Gwilliam, ex-world champion, at the forty-fifth British National Ploughing Championships in 1995

embassy that they took great exception to this degrading of Franco's crack police force, so our attendants soon went back to Ford overalls. We also had complaints that the hundreds of people who could not get into our bullring were standing on surrounding equipment, even tractor bonnets, so they could catch sight of the event. Our competitors were indeed miffed by our popularity.

In 1966 during the last winter I spent with Ford I was loaned to the service department to visit customers in the eastern counties who were experiencing problems with the new tractors, although I still reported to sales promotion and used the demonstration van. One customer I visited on a regular basis had 70 Ford tractors and was very unhappy. I was fascinated with this work and felt I could help a great deal in smoothing our dealers' path to keep the customers satisfied. Almost all the problems were simple and could have been eradicated easily if only I could have obtained some action from the factory, but I was constantly put off with stock excuses: it is not my job, we do not have a budget for that, leave it to the dealer, or we must get authority from America to do that. Although I loved visiting the customers and found the work very satisfying when action was possible, it was very frustrating when it was not possible to help as much as I thought I should be able to. Eventually I became disillusioned with banging my head against a brick wall and decided that Ford and I must part. But where would I find a job that could give me the satisfaction and adventure that Ford had provided for eight years?

8

I learned that Roadless Traction were looking for a service representative for the whole of the UK. It seemed just right for me and so I became quite keen to get the appointment. The sales manager was Mr A. V. Dodge. I had known him when he was the sales manager of the Ford dealership nearest to White City. Now he had moved to Roadless Traction and our previous contact perhaps helped in my application. One evening he telephoned to say the Directors had approved a budget for the job, so could I go to the Hounslow factory and meet them. This I did and it was agreed my new job would start as soon as the month's notice to Ford had expired.

It was with great regret that I contemplated leaving Ford, who had given me such good initial training and later experience that money could not buy (or maybe it had: *their* money). They had taken me as an uneducated country lad, sent me overseas to represent them, given me grand amounts of money to exist on, been helpful with any personal problems and without exception (including the bean counters when they were not behind their desk) had been pleasant and loyal to the company and myself. I hope and believe they had my loyalty and felt I had done a good job for them.

Roadless Traction's chief competitor was a company called County Commercial Cars, also producing, as was Roadless, four-wheel-drive tractors based on Ford tractor units. There was considerable rivalry between the companies but since both were using basic Ford units an element of contact was inevitable. At the time I joined

Roadless the main product was tractors with small driven wheels at the front and County had tractors with four equal-sized wheels. We could go on all day about the relative advantages each system had over the other but within a year of my becoming an employee of Roadless we had introduced an equal-wheel-drive tractor and County had introduced a new tractor featuring unequal wheels so I suppose each company had recognised the advantages the other's system had and, more importantly, had decided to give the customer a choice rather than try and sell him a system that might not have been ideal for his particular farm.

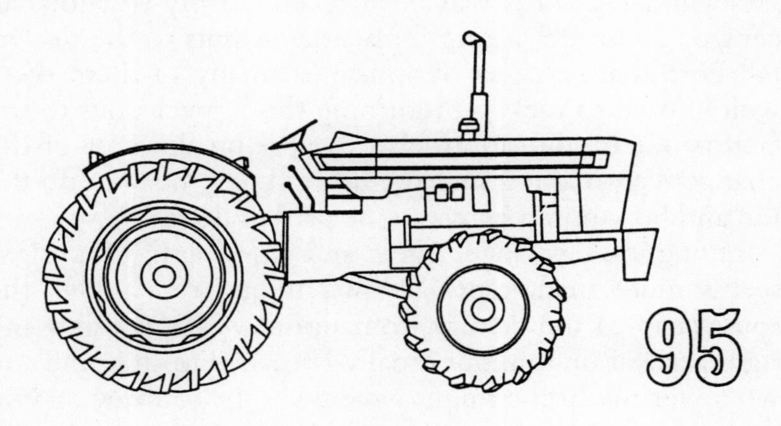

My position at Roadless was service representative for the UK. Quite some area to cover, you may think; but a new green Cortina estate car was provided and I was keenly looking forward to starting my new job, and using my new company car, the first one I had ever had. (Remember, only demonstration vans had been available to me previously.)

Shortly before my appointment Roadless had introduced a new six-cylinder version of their tractor. It now began to give some front-axle problems. Maybe the extra horsepower was the straw that broke the camel's back. The original front axle was a refugee from wartime use, basically a GMC unit and usually purchased in boxes as new or factory reconditioned axles at various surplus ex US army sales – good units but perhaps just under capacity on the six-cylinder tractor. Already about a hundred tractors had been produced and increasing breakages were becoming evident, special axle shafts were produced in Germany and a modification to the pinion bearing in the front differential was designed. I joined the company just as a decision to change all the hundred differentials on the six-cylinder tractors had been taken, so my first jobs all centred about delivering replacement units to the dealers (all Ford main dealers of course, so many of them were well known to me) and returning the suspect units to the factory for modification, also overseeing the cost of the changeover, making sure the dealer knew how to do the job and how much he was to be paid to do it.

Living in Heybridge, Essex, and working in Hounslow seems quite mad, especially when one remembers the round trip to the factory from home was 120 miles and right across London; but usually I worked from home and wherever my home might have been the journey to visit dealers would have been equally far, so the poor Cortina put in large mileages each day. There were about 170 main Ford dealers and quite a number of IH dealers on my list. The IH dealers were selling the B450 and 614 tractors fitted with a Roadless four-wheel-drive conversion so the task I had to help dealers and investigate any product problems they had was quite daunting. Monday was normally my Hounslow day. Then late in the day the journey to my first call would be started and so for the rest of the week I might be found anywhere between Kent,

Cornwall or Inverness wherever the need was greatest. Often it was cheaper to burn petrol and go home for the night than it was to stay in a hotel. I reckoned that if it was less than 120 miles to go home that was what I did. My mileage was soon averaging 1000 each week and this represented at least 20 hours work, but never counting this – so long as I could do most of the driving in the evening or early morning I was happy with the situation.

You may now begin to wonder where the character had gone who was always in different adventures, scrapes and humorous situations but you should remember my 45th birthday was looming. Was it really realistic to think my mad youth would go on for ever? Also of course I was now working almost all the time on my own and so had lost the competitive spirit of younger people leading me on. Betty and I were just as mad as ever but having dropped out of the Ford social scene and become almost a one-man band, I was now deeply conscious of being away from all authority and more or less on my honour not to waste a small company's resources.

There were still a few characters about. One of these was the works manager for Roadless, a wonderful man, Mr Skelton, with whom I worked very closely after a few weeks with the company. One of the first calls I always made on a visit to Hounslow was to see him in his office and discuss any service problems that might have cropped up. On one occasion he took a telephone call from a dealer whom I will not name and whom Mr Skelton was less than fond of. This dealer had taken delivery of a new tractor shortly before and had already made a complaint about it. After Mr Skelton picked up the telephone and found who was on the other end of the wire, the conversation went like this: 'Oh, it's you is it? What have you done to our tractor now? . . . Oh no, that does not sound as if it will be repaired under warranty. Send the broken piece back and I will look at it. . . . No, we shall not pay the labour involved if the part is not faulty.' *Click* as the receiver went down. He looked at me and smiled. 'I bet he is fuming,' he said. 'What about customer relations?' I asked. He laughed. 'I cannot go and see him,' he said. 'I will knock him down, you can go and pick him up, look at his broken part, replace it and give him as little labour cost as possible. He will think you are wonderful and Roadless also.' He was right.

Another character was Dave. He could tell me of many adventures with Roadless tractors, visiting dealers in the days when Roadless produced half-track conversions to the Fordson Major tractors and finding the tractor stuck in a field that was almost a swamp and having to wade across and try and refit the track. One of the products made was simple tracks to fit trailers, not driven tracks, just in place of wheels to give the trailer buoyancy on soft ground into which wheels would sink. During the war the army had needed a trailer to carry a tank so it had to be strong and on tracks. (Probably they needed it to recover broken-down tanks.) The requirement was for the trailer to be shipped by rail to Stockport and this was done. Now whether the tracks were faulty or the army tried to slew the trailer too sharply I do not know but a track had come off, in the middle of Stockport, blocking the tram lines. Dave was urgently despatched to see what the problem was. When you have driven the 150 or so miles from Hounslow to Stockport either the problem has gone away or you are in deep trouble. Dave was in deep trouble. The town was blocked, the trams stopped, rush hour had been horrendous but had now passed. The police were more than impatient both with the army and Dave, but the trailer was too big to do anything with so Dave produced a very big jack and started to lift the trailer, with the intention of refitting the track. The jack moved OK, but downwards, pushing the cobblestones into the ground. Then Dave had a stroke of genius: put the jack on the tramline to stop it sinking. This worked well and by the morning rush hour the trailer had been moved. But he told me that the spot was marked forever because the tram always gave a lurch as it passed over the line where the jack had been.

There were so many stories about Roadless . . . An amphibious tractor was being developed for the army, which, when ready for testing, was taken to Hounslow Heath to cross a pond. No one would risk driving it so Mr Skelton drove it himself. It sank and he had to swim. Another story was told about the Lifeboat tractor. This was a development for the Royal National Lifeboat Institution. On long, shallow beaches there were problems taking the lifeboat over the sand and finding deep enough water to launch it, so a suggestion was made to use a tractor on tracks. Roadless were approached and a Case model L tractor was tried, but spark ignition engines and sea water do not go well together and the trial was only partially successful; so someone decided that the tractor should be waterproofed. The specification that was produced called for the tractor to operate in 12 ft of water, but no one suggested how the driver was to work under water and the tractor was still a spark-ignition version. Extended intake and exhaust pipes were simple, but the sparks were very vulnerable. The engine was eventually completely cased in and first trials showed it worked. But there is always a snag. Although the engine now ran well with seawater over it, the snag was that inside the casing there was a great deal of condensation when it went cold after use and of course the magneto was inside the casing, so magneto life was quite short. More midnight oil was burnt to get over this problem and the suggestion was mooted that if the air going into the engine was directed to the engine intake via the inside of the waterproof casing it would be ventilated, thus helping the condensation problem. This was soon done; the air – cool air – now passed over the magneto and was drawn into the carburettor intake to supply the engine. 'Brilliant,' someone said . . . until one morning the engine backfired when starting and there was a big explosion. The development engineers now knew that petrol vapour and sparks do not mix very well in a confined space. The waterproof casing had been completely destroyed. After a flap valve was introduced into the carburettor intake so the tractor could not spit petrol at the magneto all was well. Roadless built many Lifeboat tractors during the next few years.

Roadless Traction had one of the dreaded bean counters but this one was different. Mr Dunworth, a man of the old school, did all his accounts in old-fashioned ledgers which were stacked on chairs, his desk and in cupboards in his office. He knew precisely how much the company spent and how much it earned each day, and could run down a column of figures quicker than anyone I knew. No computer for him; in fact the amount of cigarettes he smoked would probably have melted a computer – even the office ceiling was yellow. He was brilliant once when I was having difficulty making my expense report balance. The theory was that the cross lines and the vertical lines of figures should add up to the same total both ways. Mr Dunworth did not add them himself but told me to re-check and I would find that the figure seven or thirteen was wrong somewhere. He was correct.

This does not mean I did not have battles with him,

indeed he was taking the same attitude that the Ford bean counters had taken so I had to use the same argument that if I had to stay away from home overnight surely it followed that since my home comforts featured good food, nice surroundings, central heating, television and the attentions of a loving wife, surely he did not expect me to abandon them when forced to live away from home for the company's benefit. He smiled and retorted that he could see the argument, but there was no way he would authorise payment on one account, and that was I must not expect to be compensated if I decided to employ a stand-in wife in these hotels. With some amusement we parted good friends.

At the end of one year the Cortina had covered 64,000 miles. The tractor differential changes were completed and so I suggested that the car should be changed for a new one before it gave trouble. I had always serviced it myself, usually on Sunday morning, and never ran the engine oil more than 2000 miles before changing it. No new brake parts had been used and only one set of tyres fitted. A new Cortina Estate duly appeared and only £300 in cash changed hands. It was a good deal.

As time went on I visited Scotland more and more on service matters and eventually did sales and demonstration work as well. It truly was a long way from Essex to Scotland, so the poor Cortina covered even more miles, although the extended travel meant I could only concentrate on about six dealers there.

It really was a wonderful place to visit. Usually I was operating on the east side of the country where much of the arable land lay. There is a special type of person living in Scotland and many of the sales and service people I met at the dealers' premises were typical. We could be at war for hours over a warranty claim which I might have rejected but soon we would be in a bar or restaurant together as old friends, swapping stories about some of the eccentric customers we had.

Such as the customer who, having bought an expensive four-wheel-drive tractor, refused to take it out to the fields in wet weather in case it became dirty. Another customer with a new tractor had read all the books he could find about four-wheel-drive and had come to the conclusion it should not be engaged until the rear wheels started to spin and then only for a short distance or it might wear out the smaller front tyres. The result of this misinformation was that one day he was descending a rather steep grassy bank when the tractor started to move faster than he liked; applying the rear brakes only locked the rear wheels. He had no four-wheel-drive engaged to ensure that the front wheels were braking as well, so the tractor slid with increasing speed down the hill, crashing through the hedge, throwing the driver and leaving him stuck in the overgrowth (no cabs in those days). The tractor jumped a ditch and ended up in two pieces in the next field. The driver was unharmed but 'felt like a pincushion' was his description.

One customer decided the delivery charge for his new tractor to travel from Hounslow to Scotland was excessive so he arrived at the factory one day and just drove it home.

Often old tractors were used on the beaches to drag boats from the water. One fisherman had an old diesel Major for the job but after a couple or so years' service the front wheel feel off – this often happened because the seawater ruined the wheel bearings – so he removed the rear tyre and with the tractor stood on blocks used it as a winch, with the front of the tractor tied to a post on the promenade to prevent it from being pulled towards the sea. Eventually the post gave up and was pulled out and,

in spite of the fisherman's protestations that it was rotten anyway, he had to pay a substantial amount for the replacement post.

I well remember the salesman, senior citizen almost, who had demonstrated the Fordson tractor in the early 1920s and had been with the same dealer ever since, and the service mechanic who in his youth had driven 250 miles every day in a Ford model T truck to collect fresh fish. Then there was the story about the fitter who went to a distillery to service the tractor used to tow trailers about the yard and ran into the gate post on his way out . . . wonderful people, wonderful country.

We also had more requests for help from dealers on the Continent who had purchased used Roadless tractors from the UK. Realising that there must be a market over there for new tractors I sought permission to visit France, where most interest was, and see if it would be worth making an effort to set up some kind of operation there. So here I was starting the new strategy back at the Paris show again, not driving Dextas down steps this time but working even harder to sell Roadless. The reception was good and I took care to visit Ford France and renew acquaintance with the sales manager there, who raised no objection to my visiting his dealers looking for business. I was also trying to get him interested in taking four-wheel-drive kits from us to market through the French dealers. He already did this using Italian kits, but ours were much stronger – but as was pointed out more costly. It took nearly two years before the service problems he suffered underlined why ours cost more money and so he began to look at us in a more sympathetic manner.

In the meantime Ford decided on a new grand strategy in regard to warranty on the units Roadless were receiving from them, deciding that we must do the warranty on the entire tractor and reclaim the Ford part back from them. To instigate this scheme, which represented a big increase in warranty work for Roadless, we were invited to Ford in Basildon for a discussion and were entertained to lunch in one of the conference rooms. The people who were entertaining us were the managers who had originally been my bosses – a good feeling. It also gave me the opportunity to visit Basildon as often as I wished to discuss some claim or other.

I now also worked closely with County, so we could take a united approach to Ford in case any problems cropped up. To enable this extra work to be undertaken, Betty was pressed into service as the warranty scheme operator and it all went well. Because of my familiarity with the whole tractor, I was able to isolate some problems that were solely the Ford content and thus save myself a lot of trouble trying to find out what was wrong with our part, only to find much later the Ford content was letting us down mechanically. I could write a great deal about this period but feel it would only be a mechanic's diary, so much is glossed over. A few things, though, stick in my memory, such as receiving a call from a dealer in France who had mechanical problems and agreeing to visit him. Knowing his area was very suitable for four-wheel-drive tractors he had imported a tractor from an English dealer and could not get the performance he expected. I arrived in the south of France to find the tractor had been sold so off we went to visit the farmer. Luckily his service manager could speak a little English. We soon had the tractor in the field and to my amazement the front axle was doing nothing at all. I suppose the French dealer should have spotted the problem but it was very early in their experience with Roadless tractors. The problem was that the English dealer had sold the tractor without axle shafts in the front axle, so it was never going to work. We soon removed axle shafts from another Roadless standing in the dealer's yard and fitted them to the farmer's tractor . . . another service problem solved.

When we received another set of shafts from the factory they asked me to show them how to use the tractor to its best ability. This was right back to my demonstration days – how I enjoyed it! We set the tractor up and I made sure they knew how it should be done, and we took it to a field so it could be shown to many local farmers. It went wonderfully well and we were invited to demonstrate it to one farmer who said he would buy it if it did the job on his farm as well as it was doing that day. When we arrived he took us to a most horrendous hill, so steep the tractor would not climb it unless the plough was in the ground; then it managed it with a struggle. 'We usually work across the hill,' the farmer said; so, putting a set of disc harrows on to the tractor we set off across the slope. (Remember: no cabs at this time.) I sat on the

mudwing on the high side, ready to bale out if it seemed to be going over, but all went well and upon my inquiring whether he ever had tractors roll over he replied, 'Yes, often!' By now I was well in favour with this dealer and on my next visit managed to obtain an order for three new tractors.

Unfortunately about this time the supply of skid units from Ford began to dry up and both County and Roadless started to amass large order lists that were never caught up with. Mr Dodge, trying desperately to ration out the few units we had found, himself became totally bogged down in orders. I feel sure this was a marketing ploy on the part of Ford, who were trying to outsell Ferguson. New tractors sold as Roadless or County did not show on the market penetration figures and our sales were thus decreasing the Ford figures. If the farmer could not get a new Roadless I believe the Ford thinking was that he might buy a two-wheel-drive Ford and so make the penetration figures look better. Basildon had to answer to the Americans if Ferguson was shown to outsell Ford. . . . Back to the politics of my days at Ford.

Eventually I was able to visit France again and to my surprise obtained an order from Ford France for 300 four-wheel-drive kits to fit Ford tractors; but Hounslow decided it was too many and they would not be able to produce. This set me thinking and I began to plan marketing my own four-wheel-drive conversion. How things change – and how ambition can propel one into unknown fields!

During this spell the Cortina had been sold with 48,000 miles on the clock at only seven months old. I now had a new Zephyr 4 and it only cost the Cortina plus £100. It was now becoming plain that my service work, warranty work, sales in Scotland and also in France were causing me to spend too much time driving around. I now drove all my allotted 48-hour week so that any work done was after these hours.

Ven found a good young chap to join the company, named Mervyn Ford. At first he helped on service work and this was great for me but eventually he took over half of England as sales representative when that position became vacant; but anyway it still helped because he did most of the service work in his area as well as sales. He made good progress and now has a flourishing company of his own and I am pleased to count him still as a good friend.

Another year passed and another car went. The Zephyr 4, having done 104,000 miles, was changed for a Zephyr 6. (It could have been a Granada but that was just out so I decided to stick with a known model.) Eventually this car was sold at 15 months old with 126,000 miles covered. Green Shield stamps were being given away with petrol at this time. Sometimes one would see the sign 'treble stamps today'; I reckon I had my share, and Betty enjoyed choosing the goods from the catalogue.

I remember our dealer in Toulouse with particular affection. On one occasion I was asked to visit a farmer who had bought a County tractor only to find it was broken down. The farmer demanded I give advice on the repair and in spite of my insistence that we did not manufacture that machine he appealed for help. The dealer was, of course, present so a round table conference was held and the main problem isolated – in any case it was a problem known to me. But spare parts were going to be hard to get, it being a Ford part that was giving trouble, and the replacement cost was too much for the farmer to contemplate. In the meantime the dealer had taken advantage of the situation and suggested he buy back the County and replace it with a Roadless. An agreement was reached that if I would instruct on the use of the Roadless the farmer would change tractors, so late the next afternoon, we arrived with our tractor and the farmer brought his other County tractor to compare with ours. Again we come to people with individuality; our French farmer arrived wearing bedroom slippers, and I remarked on this unusual footwear. 'Oh, yes,' said the dealer, 'he always drives his tractors and the combine in slippers; he never gets off the tractor.'

I was well prepared. Our tractor having smaller front wheels, it was likely to have less traction but the County was not well set up and so we were able to at least equal its performance. Afterwards we were invited into the farmhouse for the evening meal. Did I say farmhouse? It was a monster building in the form of a square with a central courtyard where all the machinery was kept.

Between the twin towers at the front was a gateway, giving the impression of a castle, but made from brick. We were soon being treated to an Armagnac apéritif. The father made this potent brew himself and I was taken to see the cellar where he kept it. I had the impression that he had enough to power a fleet of Lancaster bombers and I reckon they would have run on it. Much wine, wonderful food – and later at around midnight I made the hotel.

Another great character we had as a dealer was a man just north of Paris, not quite as staid as many dealers I knew, but he was a definite man of passion when it came to business. I was taken to his home and became one of the family. We originally met at the Smithfield Show in London where I had agreed to introduce him to a few English dealers who might have used Roadless tractors for sale. We met after the show and I entertained him with Betty at home for a meal. We always got along great and later he took a small amount of equipment from Roadless. One day whilst I was trying to persuade him to take a stock of tractors, he decided we should visit the motor show in Paris. In keeping with his character he told his spare parts manager to act as driver. (The poor man had no choice but I am led to believe he was used to this call to duty.)

We arrived in Paris and spent much time talking to people he knew and taking champagne with them. After buying several sets of tools and a stock of oil and an initial discussion about importing an English kit car to France we left the show. Now the dealer decided he would drive. This gave me some concern – it seemed obvious to me he was not in a fit state to do this, but he insisted. All went well until a set of traffic lights turned to red and a small car in front stopped and we did not, at least not quickly enough. There was silence after the initial impact, which was not too great; then the irate driver of the front car got out, looked at the rear of his car and came to the driver's window of our car, obviously expecting to engage our driver in conversation. But the window remained firmly up, our driver staring straight ahead. A tap on the window produced no movement, nor did a voluble flow of French, and by now a rising scream of hooters told us the lights had changed to green. Much more use of the French language, probably oaths, produced nothing and so the driver climbed into his car and drove away. This must have made our driver thirsty, or maybe he was going to a nightclub anyway, because that is where we went. More champagne and about 3.30 am we arrived at the dealer's home with the spare parts manager driving. My car was locked up in a security compound and so I was blearily told to take the Granada belonging to the dealer to my hotel. I returned to England the next day.

But travel was hard work; in fact tractor business was hard work. I visited this dealer only once more, but by then the demand for tractors in England had snowballed to such an extent that it was impossible to meet the need. Shortly after this last visit I was told my friend had been killed by a train whilst out shooting. A good friend had gone leaving some wonderful memories; a tragedy indeed.

A company vaguely connected with Roadless was one based in Turin, Italy, by the name of Selene. There was an exchange of parts and ideas but the main connection was the holding of patents and using them. Selene held a patent on a sandwich-type drivebox to fit various makes of tractor, the same drivebox as used by Roadless for all the Ford 4WD conversions that they produced. Although the box itself was made by Roadless, the system was covered by the Selene patent and thus a small royalty was due to Selene for the use of this patent; also Roadless had an agreement not to sell units in Italy and Selene did not sell units in England. On a few occasions I visited the Selene factory in Turin and found the owner, Mr Segre Amar, an agreeable and charming man. (A few more miles for the Zephyr to cover.) In addition some equipment was imported from Holland and Germany, so these were further destinations to visit.

In the fullness of time we were notified that Mr Segre Amar had retired and that Selene had been sold to a Swiss company called Schindler. This was just prior to the annual Paris Show. During the visit to the show I was accompanied by the Roadless chief engineer and of course part of our job was to look at all the 4WD units available and assess them for comparison to those produced by Roadless. Naturally we visited the Selene Schindler stand where we met Segre Amar Jun., who was now the sales manager of the agricultural department of Schindler, and Mr Wild, the director of that department. I was to have a close connection with these gentlemen in the future but at that time I did not know this. The main interest of the visit was an axle displayed on the Schindler stand that had several features quite new to my companion, leading him to believe the unit on show was much in advance of anything he had seen before. We both thought that if it was to be imported into England it would offer very keen competition indeed. After discussion we decided to make sure that if anyone imported this axle it should be Roadless Traction, and so we requested Schindler to extend the facility we enjoyed with Selene and made a verbal agreement on the spot that Roadless should have the UK import right to this equipment. Of course we did not have the authority to make such an agreement but felt it necessary to protect the company's interests in the short term whilst the directors of Roadless made a final decision and signed a binding agreement if they wished to do so.

As a result of this agreement it was indicated to Schindler that Roadless would take 60 units in the next year and an official order would follow. About five months later I received a telephone call from Schindler to say that the order for the 60 units had never been received and could I do anything to help. In fact it was not my job, but agreeing to try to help I approached Mr Booth, the managing director of Roadless, a wonderful man who in the past had given me a great deal of support. He indicated that Roadless had decided to produce their own new type axle. (I felt that price might have been behind that decision.) He went on to say he felt rather badly about not informing Schindler of this decision but perhaps I might do so verbally whilst he would confirm in writing shortly. In my opinion the Schindler axle was far in advance to anything available in the UK and although there was certainly a price problem it seemed to me it could be overcome. Of course I was authorised to talk to Schindler, and almost as an afterthought I said to Mr Booth that perhaps I would import the axle. What surprised me was his reply; he said that had to be my decision . . . but did not go on to say that they would be sorry to lose me – or even good riddance.

9

NOW before we go on with this part of the story it is necessary to turn back almost two years. The death of my mother had caused us to leave Essex and return to live with my father; so again we were living in Ambaston, the place of my birth. Our return to the farmhouse that had been Betty's and my first married home brought back many good memories. The old farmhouse had been offered to my father at a rather silly price because he was the sitting tenant, so when we tragically lost my mother we decided to move house.

My father was on good terms with a local builder and we soon made friends with him. He was yet another character, one of his own. Born to relative poverty in a poor part of Derby he could tell me of the days when he was 12 or 13 years old and worked on the open market in the town to earn sixpence on a Saturday morning so he could pay his admittance fee to watch Derby County play football at the baseball ground in the afternoon. His name, was Bill. Unfortunately I have lost touch with him but if he reads this he will know who I am writing about and will surely remember telling me his story, how having once managed to save the astronomical sum of 5 shillings he decided to travel to an away match to support his beloved Derby County. So he could appear presentable on the coach he borrowed his brother's suit, but omitted to tell him. His 5 shillings paid the coach fare, entrance to the ground, and bought a portion of chips on the way home. Unfortunately he was not able to replace the suit before his brother found out it was missing and so had to endure a thrashing. He told me that sometime during this era he decided he would own a Rolls-Royce when he was thirty years old – his notion of being an old man. The Rolls-Royce came two years late . . . an even older man I guess.

On one occasion I persuaded Bill and his wife to visit the Royal Highland show and stay for a weekend break afterwards; so as soon as Betty and I could clear the show, off we went in the Roller, as we called it. I had booked us into a rather classy hotel at Pitlochry. Up a tree-lined drive to the rather imposing portal . . . no-parking signs around it and a polite request to park in the car park. Bill drew up right on the no-parking lines and a porter emerged from the portal to collect the luggage. No word was mentioned about the no-parking notice; you see, no one will ever mention it to a Rolls-Royce owner, Bill said. He was right: no one ever did.

We had a lovely weekend, deciding on Saturday night that the rather swish hotel did not offer the right ambience in the bar for a night's working-lads' drinking. We went off in the Roller to look for a pub, finding one a few miles up the road. Bill gave me the Roller key. 'You drive home,' he said, 'I intend to enjoy my drink.' I enjoyed mine, uncountable bottles of tomato juice. About 11 pm the barman asked whether we wanted to stay longer, as it was closing time. 'If it is possible,' we replied. 'Here are two room keys. If the police check, show them you are residents,' he said. When about 2 am the police arrived, they were welcomed as friends with a lager and a dram as a chaser to keep the cold out. Eventually they departed, no doubt in search of lawbreakers like ourselves. They never asked if we were residents, but of course they knew we were not. At 3 am the car park was empty. We had all gone home, the landlord having done his best to persuade us to return for Hogmanay.

Eventually we decided to have a new bungalow built, and Bill and I began to go out together quite a lot, often talking about the tractor market in the UK. In short, Bill and I decided to start importing an axle from Italy and managed to have an engineer produce us a design for a drivebox which we would have manufactured. I told Roadless about this decision and agreed with them that we would only market the kit in a market where Roadless did not have a model. So we had a company, Farm Tractor Drives UK Ltd.

All the problems we had will not be listed here or this book might seem like a bank robber's alibi but eventually we managed to get a product together that was saleable. My contacts with the Forestry Commission led to the sale of a unit on a trial basis to them and eventually we found a fork-lift-truck manufacturer who took units from us; but in due course he went into receivership and was sold, so that customer disappeared. During this period my work for Roadless carried on and the extra work for Farm Tractor Drives was mostly undertaken at weekends but it was inevitable that there would be some overlap.

As this book is written it brings back many memories. For instance I had already bought a Fordson tractor of 1933 vintage because the interest in collecting old tractors was just starting and during my visits to a Roadless customer I saw the old tractor in a bed of nettles. Inquiring about it I was told that the farmer had bought it new and still owned it. Would he sell it? I asked. After some thought he said I could have it for £30 if I would not sell it on for a profit. Indeed I would not – it was my intention to restore the tractor and use it. The main snag was that it stood in the nettles in Norfolk and that was the best part of a long way from Derbyshire. But when I told one of my friends about the tractor, his response was, 'Find me one out there and we will fetch both of them on my trailer.' On my next visit the local dealer told me of a scrapyard where many old tractors lived; indeed there was an old Fordson in there and I arranged to buy it for £40. Did I want a David Brown? I was asked. It lay in a ditch on its side but still had a good coat of paint and looked in fine condition – at least it was all there. I refused it partly because he asked £70 for it and partly because it was too new. (It was a 50D. . . . I must have been mad: it would probably be worth several thousand pounds today.)

On the appointed day we set off for Norfolk with a petrol Land Rover pulling his rather long trailer. We left home about 6 am but didn't travel very fast. Having collected the two tractors we set off on the return journey around 4 pm. We now had about two and a half tons of old iron on the trailer plus the weight of the trailer itself and the petrol Land Rover was objecting. Its way of doing this was to lose control of the trailer at speeds in excess of 25 mph; in other words we were swinging all over the road – most dangerous – but we made it home at about 10 pm, cold, miserable and wishing old tractors in a very hot place.

Our old Fordson proved very useful. It removed the hedges that were in the way of our new bungalow and attended some old tractor rallies even whilst in its rusty condition. Nicholas and I rebuilt the engine and had it running very well. It was whilst doing this repair that I started to call on scrapyards as I travelled around the country in search of spares. Soon my friends who looked at the tractor as we repaired it began to say, 'I could do with one of those so my old Fordson can be brought back to life.' I now started to collect as many parts as possible. This soon led to selling them on a stall at the various steam rallies . . . and another business had begun.

You may think I was greedy to keep developing all these differing types of companies but really I had no intention of growing so quickly – the ball was rolling and it just had to be chased. To recap, we had: Farm Tractor Drives selling four-wheel-drive conversions to fit Ford 4000 tractors, Tracprez Parts selling vintage tractor parts through the post and at steam rallies, Tracprez Publications, which was now just starting and was intended to produce and sell service and workshop manuals for vintage tractors (this included *Fordson Magazine*, which was meant to be a vehicle in which we could publicise our spare parts) and also the workshop manuals and of course I had a full-time job with Roadless Traction as service manager covering every Ford tractor dealer in the UK. I also visited the Irish Republic and had responsibilities in France. Even the driving part of the Roadless job occupied more than a working week, close on 100,000 miles a year; at that time my estimate of a working week was at least 100 hours and some people complain if they have to work a 40-hour week. I enjoyed every minute of my week.

When I became involved with vintage tractors I could see the need for a supply of parts, so not only did I scour the scrapyards for parts but realised it would soon be necessary to produce purpose-made ones and to this end several small companies were approached to see if they would undertake the manufacture. This meant of course that finance had to be found, and from this need *Fordson Magazine* was developed. If you have parts to sell, which we had in increasing quantities, the answer (so the experts tell us) is to advertise. Advertising is a very skilled business and I was not sure it was within my capabilities to secure the best at a reasonable price. My way was to offer the vintage tractor users an interesting magazine which they would pay for, but would give us the platform to advertise our parts and services at no cost to us, and be sure the advertisements were reaching the right people. In return our subscribers had good technical articles on repairing their machines and the availability of parts to use on them.

The technical articles were meant to be backed up by an offer of parts to do the job; not always possible to offer everything but at least I could tell our subscribers how to do it.

A low-cost printer was the first need. We found him and Ford Tractors were put under light pressure to buy a full-page advertisement, as were Ransomes. Both companies came up trumps and this paid our first printing bill. We printed 500 copies of our first specimen issue and before we realised it they were all sold. We never printed less than 1000 after that and were soon printing 2500. I was writing articles and stories to put in the magazine even in hotel bedrooms at night.

I joined the National Vintage Tractor & Engine Club about this time and when the editor of their magazine resigned, we produced, edited, printed and circulated that publication for a short time until its present editor, Brian Sims, took over. Betty was deeply involved of course, still doing the office work associated with Roadless warranty, helping on the stall at weekends, typing etc. for *Fordson Magazine* and acting as a bean counter for Farm Tractor Drives. Even Nicholas was involved in various ways. We were a very busy family.

Our axle we imported now produced another surprise. A French company used it on a fork-lift truck and began to import it into Britain. We had sole import rights to it but not in this case because it was being brought in as part of a vehicle, but we negotiated an arrangement where we drew a small commission and in return gave service and warranty back-up to the axle and provided a parts service. The agricultural machinery trade was a very complicated one, and this was a typical example. The French fork-lift truck using our axle was driven by a drivebox designed and produced by Roadless; I was required to give service instruction courses on the drivebox and follow those with a course on the axle and do warranty on the drivebox for Roadless and also for the axle, which had to be claimed from Sige, the axle supplier in Milan, Italy.

Mention of Milan reminds me that when we first imported the axle, five were sent to us by lorry. We had no commercial transport of our own so I decided to purchase a transit van, both for axle carriage and to take our stall around the rallies at the weekend. An urgent demand for axles meant we could not obtain them quickly enough by lorry, so Betty and I decided to collect some from Milan. A quick oil change on the transit and off we went. After a lovely ride through France and Switzerland, through the 30-kilometre Mont Blanc tunnel and into Milan, we loaded our axles and returned to the valley of Aosta to stay overnight. The next morning as we set off heading for the tunnel there was the most terrible noise from the rear of the transit. I ran with as little power as possible and the noise was not so bad and – yes – we made it through the tunnel (and, I might add, with some relief – it certainly would have caused consternation if they had found their tunnel blocked by a transit van). Thinking about our problem I decided to check the rear axle oil level. There was no oil in it! And I am supposed to be the service man who reads the riot act to dealers who treat our equipment like this. We put some oil in it and the noise was never heard again.

Perhaps we should take a closer look at the old transit

van that took us to Italy. It had one of the 'V' 4 engines and must have been quite an early model. It was bought from a nearby garage-owner who had repainted it. He said it was a good van. (Well, he would of course.) We paid £150 for it and it covered many, many miles including being run without rear axle oil in Italy. The engine never used any oil.

It became well known on the steam-rally circuit. We were soon visiting events from Dumfries in Scotland down to Liskeard in Cornwall and at that time were selling a number of used spares as well as the new ones we were producing. I well remember going to one event because the van was so heavily loaded that near Hungerford one hill proved so difficult for the van that I had to progressively change gear until we were in first and losing revs, but just made it to the top. We had as part of our load a used Fordson gearbox, the high top gear variety. This was soon sold on the site to a man who said his Fordson was so slow that it had taken him eight hours to reach the site. The last we saw of the gearbox was in a wheelbarrow heading for the 'slow' tractor, after which the man said he was going to buy a set of pulley blocks he had seen in the autojumble so he could use a tree branch to fit the gearbox in his tractor before the journey home. Some people are really keen.

At that time we did not have a caravan but slept in the back of the van, so when we arrived on site there was some urgency to erect the tarpaulin cover over the back of the van and set up the stall so we could prepare our bedroom. The inside of the van was draped in muslin and this coupled with the blown-up bed and luxury sleeping bag led to our transit being called various things by the customers, who could plainly see inside the van when the rear doors were open. Some of our customers had various types of humour. 'Boudoir' was a more polite description but 'mobile brothel' and 'knocking shop on wheels' were some of the others I can remember.

We always had the rear doors open because we needed to sit on the back of the transit with the stall in front of us. At night we shut the front of the stall, i.e. folded the spare piece of tarpaulin over the front of the stall, and we had a small area where we could sit and do some elementary cooking with a small gas burner. The transit bedroom was too clean to go in until bed time, but we had plenty of home comforts, like a few bottles with various liquids in them, a rough old carpet to cover the ground and chairs to sit on and often visitors to yarn with. One rally was so wet; it rained and rained; during the evening we sat talking to friends, periodically lifting the top of the tarpaulin to run off the water until after dark we perceived that a small river was running through our 'house', at which stage we retired to bed. At one Kendal rally it rained on the journey, all night, all the next day and the following night, so we could sell nothing; but about midday on Sunday it cleared and the sun shone. Good, we will unload. We had not slept in the van the night before, taking B&B in the town. As we began to unload, people came and bought almost everything as we took it out of the van. We sold as much in half a day as we would have during an entire weekend.

Now we must return to the Schindler problem. I did as Mr Booth asked and contacted Switzerland, telling them that Roadless would not proceed with the verbal order. They were not well pleased with this situation because already some parts had been produced in anticipation of the order coming. I suggested that a visit from me might help as there were certain ideas that could be discussed. Mr Wild intimated he would be glad to see me and a week later I flew into Basle for the meeting. Mr Segre Amar Jun. met me at the airport and took me to the Schindler factory in Pratteln, just outside the city. A factory tour followed, a presentation telling me that Schindler had 50,000 employees in 23 factories throughout the world, the main products being railway rolling stock, tramcars, elevators and escalators. It was most impressive; so much so that I began to wonder what I was getting myself into this time – small-time mechanic, tractor demonstrator etc. – could I possibly handle what this international company would demand of me? After lunch, at a meeting with Mr Wild, we went over the Roadless situation. They understood that a manufacturer such as Roadless would prefer to produce their own unit. But what about the UK market? Could I suggest another importer? They knew of my involvement with the Italian axle. Mr Segre Amar (recently involved in the Italian market) was familiar with the situation.

I suggested perhaps Farm Tractor Drives could handle the product but finance would be needed. Mr Wild, much to my surprise, offered me three months' credit on ten axles. All I had to do was find £22,000 by the end of that period; also a stock of spares would be needed if the project was to be correctly handled. It frightened me to death. My co-founder and partner in Farm Tractor Drives, Bill, had already pulled out, and Betty and I were the only shareholders. As of that day it is doubtful if we could have paid for even one axle. However it has never been my policy to chicken out; a great deal of research had been done by me over the previous week and I was certain the Schindler kit would sell well in the UK.

The axle deal was offered to me and Schindler agreed they would supply no one else in the UK, unless a mainline manufacturer wanted to take axles for his production and was prepared to order in excess of 50 units, in which case I might be called upon to act as Schindler's representative in the UK in return for having my expenses refunded. It was suggested I might like to think over the project and let him know in a few days. 'No,' I said. 'I will accept now.' 'OK. Can we put back your flight and take dinner together tonight so the details can be discussed? Tomorrow we can sign an agreement.' 'So long as the agreement is simple and I can understand it,' was my reply.

So the next day I returned home with a deal that was to prove very profitable. We were now in deep water indeed; we had ten units on the way from Switzerland and an old chapel as our headquarters (rented I might add). It seemed obvious that I must now give up my job with Roadless as indeed must Betty. We had limited capital, £22,000 to find in three months and must publicise the units to sell them, our only vehicle an old transit van.

Betty and I decided to find out how much money we could raise from our various accounts and how much overdraft facility we could call upon. The next day I reported to Roadless Traction and gave an account of my visit to Switzerland. To my amazement Mr Booth was very concerned, not, as I might have expected, because he now had another competitor in the market – even if it was indeed of small resources – but on account of my own wellbeing. He felt we would not sell ten units in six months, never mind three months, and suggested I wait a few weeks before resigning to see how things worked out. 'You may need the job,' he said. But really I regarded myself as being in competition with Roadless and told him I felt it was unethical. 'Just wait a few weeks,' he insisted. My colleagues at Roadless treated me as if nothing had changed, so at least for a few weeks I still had a job.

Before I had visited Switzerland, Mr Tiplady, the tractor manager at Ford, had been made aware of my visit and what I hoped to do. During my period at Ford he had been my boss for a time and I had a great respect for him. His reaction was that it was OK to sell to the Ford dealers but 'Remember we may in the future decide the market is large enough to introduce our own 4WD tractor, so be very careful you do not get your fingers burnt.'

Looking back now, it is my belief (although there is no evidence to back it) that because there was a market-penetration war going on between Ford and Massey Ferguson to see who could call themselves market leader, the introduction of a four-wheel-drive kit to the Ford dealers suited Ford. You see, the tractors sold by Roadless, County and Muirhill, although of Ford origin, did not show as Ford tractors in the registration statistics. That is why in my opinion the supply of skid units to these independent manufacturers was not always freely available; because it might well have been thought that if the customer could not obtain a four-wheel-drive unit from those manufacturers he could easily decide to take a normal Ford tractor from his dealer, thus helping to increase market penetration for Ford. Perhaps now our Schindler kit could give the Ford dealer an edge over his competitors, because the tractor would still be sold as a Ford tractor. That is why my feelings were mixed regarding Roadless; I felt, as time went on, that I was offering more and more competition to them and felt very uneasy about still working for them.

Our bank manager was helpful, allowing us an overdraft facility of £5000, but even so required the deeds to our bungalow to be deposited with him. But still we had to find literature, better office facilities, a basic stock of Schindler spare parts and shortly would have carriage charges on the axles to pay. Life was getting complicated indeed. We had to provide fitting instructions in English, driver's instruction books and spare parts lists, so all our time was well filled. There was no time whatsoever for social life. Of course a complete list of Ford dealers was available to us and within a day or so each dealer had a letter announcing our appointment as Schindler importers. Each day I visited several dealers as part of my Roadless duties and they soon wanted to talk about 4WD kits more than Roadless service and warranty. It was difficult but even now I feel satisfaction that I never neglected Roadless duties to concentrate on our own venture, but it was never easy.

After our introductory letter had been sent out and a few days later a leaflet and fitting instructions I expected interest to be shown, because I knew the current 4WD manufacturers were only scratching the surface of the requirement for these tractors. Three days after our leaflet went out our telephone was very busy indeed with dealers seeking more information and – thank goodness – actually placing orders. By the end of that week (our first full week's trading) we had sold all our ten units and were on the telephone to Switzerland for twenty more.

Looking back, I cannot see how we ever coped with all this work. It was OK to select an axle ready for a dealer, and a drivebox was not too difficult, but each one had to match the gearing of the tractor it was going to be fitted to. Our main problem was the mass of bolts, brackets, and assorted bits that came in a large crate, leaving us to ensure each dealer had everything needed to fit his kit. This was quite difficult . . . but we were committed, there was no turning back now. And I was the delivery driver. So often a kit would leave our chapel after tea, overnight stay in a hotel, deliver kit early next morning and then carry on with my Roadless work for the rest of the day.

Luckily Nicholas was nearing the age to have a driving licence and just finishing a year at agricultural college. We had planned for him to spend at least a year away from our business, but that looked less likely now and, sure enough, having finished college and passed his driving test he became our delivery driver . . . and soon the person to instruct dealers on kit fitting, travelling all over the UK.

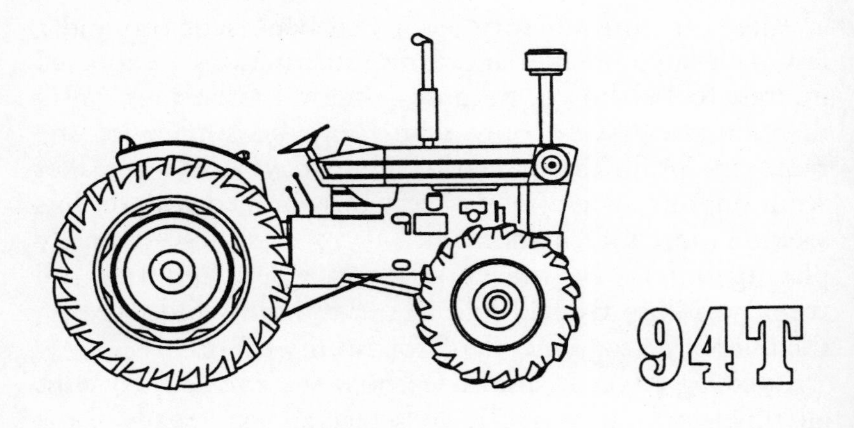

10

AFTER three months of this we were approaching the Smithfield Show period and again I raised the subject of leaving Roadless but again was asked to stay at least over the show period. It was so difficult having Roadless service to talk about on the showstand and having many dealers wanting to talk about Schindler kits. I felt very embarrassed about this and finally left Roadless at the end of December. We changed the transit van for a more modern one and I bought from Roadless the car that I had been using.

Our money problems were over, at least for the time being. Now instead of travelling for Roadless I could visit Switzerland quite often; the boot was now on the other foot. It was not 'How many units can you sell?' but 'How many units can we have?' Our market was fantastic. We worked out a delivery schedule for the next 12 months requiring us to place orders for 120 units and committing us to spend a quarter of a million pounds. It felt like winning the pools but always I was able to remember my small beginnings and tried very hard to steer clear of financial problems. We were able to enlist the services of a young lady from a nearby village as our first employee. Perhaps that is not quite accurate; we had a very pleasant and capable young lady for a time previous to this but transport problems caused her to leave, so Hazel came to replace her. Soon Nick's girlfriend came to help with our financial accounts and our wonderful association with Michèle as employee, girlfriend and eventually wife to Nick had started. We were also to enrol Janet, a very good typist and wonderful for dealing with customers, knowing just how far to go with a small discount or other assistance to persuade a dealer to take two units instead of one, etc.

In the meantime my visits to Switzerland were quite frequent. Much effort was required to look after our warranty system; articles were needed for *Fordson Magazine* and quite a number of spares for old Fordson tractors had to be commissioned from various machine shops to keep our supply of parts viable for the steam rallies we were still visiting in the summer weekends. Nick had chosen a new VW pick-up truck for his deliveries – he almost lived in it. We were doing quite well.

One of the rules I always tried to operate was that if a customer wanted something it must be available *now*, not next week or in the future but *now*, and so often we could receive an enquiry for a 4WD kit in the morning and could say to the customer, 'If you have the cheque ready this afternoon it will be there.' Usually it was no later than the next day. This helped our sales in no small measure. Our customers could feel confident in selling a tractor, knowing that the part they needed from us was almost certainly assured.

Until I started to write this story perhaps my conception of how far we had progressed in a few months had not been clear to me but the realisation that by this time we had a new Granada estate car, extra staff, new pick-up truck and – most important of all – were no longer operating on the bank overdraft but financing these deals ourselves is telling me something I had not previously grasped. Our accountant (yes, we had a tame bean counter and very valuable he was) was horrified to find we were not borrowing money as a tax-saving move; but I reckoned borrowed money is bad money and resolutely refused to borrow if it could possibly be avoided. I guess many companies at that time were wishing their life was that simple. In addition to Ford Tractor Drives we had two extra limited companies, Tracprez Publications and Tracprez Parts Ltd, and another was projected. This came about due to visiting the Verona fair in Italy. If you are responsible for a company you must keep moving ahead. This company, registered as Cultient Ltd, was meant to import small grass mowers and garden cultivators. My idea was to buy at low cost, selling direct to the customer, whilst at the same time offering a rental service whereby, for a fixed yearly sum, someone who worked full time and wanted to do his own garden could be sure his cultivator or mower was always available. If a machine failed we would be there with a replacement as part of his yearly hire deal – and very quickly, so our hire service initially would operate within 40 miles of Derby. Unfortunately although I offered several people the opportunity to become managing director of this projected venture no one seemed to want it. In today's climate maybe it would have been different.

Our accountant was still going up the wall about the amount of tax we were paying and wanted us to increase our parts stock and borrow money to do it so we would pay less tax. At that time there were tax concessions on stock; but I knew that one day we must sell the stock and then have to meet the tax bill . . . and how were we to know that it would be as easy to pay the tax then as it was now? I could not convince our bean counter that my ambition had always been to pay a great deal of tax, because if one pays tax surely there is also a lot left. By and large this has proved a good theory so long as one was able to take advantage of all the little legitimate dodges the Chancellor usually provides (because he is a bean counter and usually a little naïve in financial matters), but I must say they are getting a bit more crafty in the 1990s.

Whilst I am writing about banks there is a true story worth repeating about the time we decided to have a look at the export market for our kits. The United States seemed right for various reasons, one being that there was no 4WD conversion available to the Ford dealers over there. I flew out and talked to a few people who were all keen to import our kit, but when we talked about prices the best deal that was on offer would have meant that the kit would sell for one third the price of a new tractor. The importer would not operate at less than 120 per cent of add-on cost and one firm I spoke to felt the basic price we offered would have to double to make it worth their efforts.

It only needed two days to tell me that having an importer was no good to us but maybe Schindler might be able to import direct. But they were not too interested, so finally I decided to talk to a Ford dealer and selected one who had already shown interest in the product. He agreed that it must be a good deal at the price I quoted – but what about delivery? It had to be quick, but we could put kits into a central store and when we received money wired by the dealer from his bank to ours we could authorise the store to release the unit. That worked well but was tiresome in the paperwork required.

Our bank manager became interested in this project; our bank having connections with an American bank, he suggested he ask his export specialist to call and see me. He arrived at the appointed hour, pleasant man, nicely dressed but about as useful as a spare groom at a wedding. He was able to map out a whole strategy of how to export goods, how they could help us to find an importer, how they could help with customs and other paperwork for a small fee, how they could arrange transport for a small fee, and arrange for the necessary overdraft facilities that we would obviously require. In short he spoke the biggest load of cobblers I had heard in many a long day. He obviously had not done his homework or he should have known from our account that it was unlikely we would need a loan, and when I told him my market research had proved his theory about an importer had been found wanting he seemed speechless. I was not very polite but saw no reason to be so; the man did not know his job as well as I did. So much for the bank's assistance.

At this time we were really overloading our chapel. The office space we had provided on the upper floor, where the choir may have stood, was quite adequate but the lower wooden floor must have been desperately overloaded with all the cast iron chunks we were storing on it. Bigger premises were becoming the pressing daily need. Around the corner from our chapel was an old warehouse recently used as a grain drier by a now defunct corn merchant. It was terribly damp and the brickwork was crumbling but it was big and very attractive because it was close to our office facility. The main snag was the roof, tropical type – one could see the sun through the rotting rafters. The owner was keen to let it but realised it was in no state to use at present and he could not afford the money to put matters right. It was jointly owned by a sheet-metal-working company and a builder. The sheet-metal company needed all its capital for expansion and the builder was using his capital and no doubt more besides for development.

I looked at the situation and decided a typical Arthur Battelle deal was needed, so discussing the situation I suggested that if we paid for the roof repairs we should have the building rent free for a period. The estimate for a new roof was £10,000. OK, we have it for five years rent free and have first refusal on the next five years at a rent to be discussed at that time. It seemed good to me but having paid the first £5000 our bean counter went mad. 'Suppose they go bankrupt?' he said. 'Well I guess we would have to buy the rest of the buildings and the sheet-metal company,' I said.

Our new warehouse (known as the Iron Warehouse) had a barge quay at the end of it where cannonballs had been brought by canal barge and later as they were required were transported to the east coast by canal and the river Trent. I thought that was why it was called the Iron Warehouse but as I began to learn some of the history of the building, I discovered there were cast iron brackets, big ones, supporting an upper floor that partly covered the area and these brackets were very early examples of cast iron being used for this purpose. There were no bolts to fasten them together; they were locked together and held in place by pegs just as wooden beams would have been fastened. Several historians came to look at them during the time we used the premises and I only hope they survive this modern world.

When we had moved into our new store we began to talk to a man living on a canal narrow boat which seemed permanently moored to the quay near our premises. We soon found he was a good mechanic and he proved very useful to us over the next few years. When our fork-lift truck arrived he must have seen this as an opportunity to clear a problem he had suffered all the time he lived at that place.

One day we found him in bathing shorts by his boat and in a fog of muddy water delving into the depths. His explanation was that for several years every time he drew the boat up to the quay it hit something hard underwater.

He had believed it to be a big stone, but now found it was metallic and so had hitched a rope to it. Could we lift it out for him with our fork-lift truck? Gingerly approaching the dock wall, not wishing to find out if our truck was amphibious, he tied his rope to the forks and an old engine was lifted out of its watery grave. We assumed some boat owner had decided to fit a better engine and just chucked the old one over the side. I immediately recognised the engine, an old Ford model T.

We could now use our newly covered store, so paid the second £5000 and moved in. The fork-lift truck came and our faithful Fergie 20 with loader. The only thing for moving axles that was low enough to go in our chapel having been sold, we now found the owner of the chapel prepared to sell it; and so we had a good office which was now our own and a warehouse with a sure five year tenancy, we had a good vehicle set-up, a good market and our turnover was approaching a (to me) staggering half million pounds a year; we had a showstand, albeit small, at the Royal Highland Show, Royal Show and Royal Smithfield Show.

It was like old times visiting the shows again and meeting many old friends. Betty and I usually worked until 7 or 8 pm and had dinner at a restaurant on the way home and although we were working a seven-day week, with writing and visiting steam rallies, it was never an oppressive time. We enjoyed our life very much. I took great pride in progressing from a country boy earning 8 shillings a week in 1938 to having a lifestyle allowing me to fly around Europe, drive a new Ford Granada car and live in comfort in the early 1980s. But we knew there were only about two years of life left in our kit market.

Now flying to Switzerland quite regularly to discuss increasing our schedule of axle deliveries and sort out what few warranty claims we had, I was developing a good relationship with Mr Wild of the Schindler Company. On one of these visits he asked me to work for them as their representative for all English-speaking countries, just for the agricultural market. 'It is impossible,' I said. 'Look at the mass of equipment we sell in the UK.' 'But all is going well over there,' he said, 'and we shall not often need you.' After some discussion over lunch I had a new job, unpaid but all expenses reimbursed. The Schindler equipment was building a good reputation for quality, not just in UK but throughout Europe as well, so I did not expect to be called upon too often. Taking on this extra work I had thought back to how helpful Schindler had been to us in starting up our UK market and felt I should now have the chance to repay them a little.

The first call upon my services came in February. It was to Helsinki in Finland. 'But that is not an English-speaking country,' I said. 'Yes,' was the reply, 'but you find anyone who can speak Finnish, and they all speak English.'

As always, I first travelled to Amsterdam. Because we live so near East Midlands airport and there is a good service to Amsterdam I always used it in preference to Heathrow because that airport was much more difficult for me to travel to and the aircraft on the Amsterdam run were Viscounts – to me almost as good as the old DC3 that used to be around. One of two seats at the back of the craft had extra leg room and the hostess always had her travel bag on one seat but if you made the effort to get on the aircraft among the early boarders you could pop down on the other seat at the back whilst the other passengers rushed forward. Why they should do so I do not know, unless it is this brainwashing that many companies' personnel training courses do to inculcate the 'get up and go' syndrome, making them wish to be first in everything.

(Most people travelling on the early morning flights were business travellers.) Once the aircraft had taken off the air hostess was so busy she never came back to this seat and so one had two seats to oneself until it was time to land.

There were three calls to make in Finland: Ford, Massey Ferguson and a machinery manufacturer about a hundred miles north of Helsinki. I was taken to see this manufacturer by a representative of Massey Ferguson. We drove north through deep snow. In Finland they do not seem to have snow problems, six inches makes no difference to the traffic, or so it seemed to me; there is simply superb equipment for dealing with it, and the snow is dry, not wet and this seems to help in its clearance. We drove through snow banks piled up at the side of the road as high as six feet and with temperatures as low as 25 degrees of frost – lower at night. In one place we turned off the twisty road and drove at 60 mph across a lake, following cones in case of fog. Meeting us was a Volvo articulated lorry carrying logs. It must have weighed all of 50 tons,

skid chains, hydraulic crane and all. 'How deep is this lake?' I asked. 'Oh about 30 metres, I think,' was the reply. 'Is the ice thick enough?' I asked, getting worried by this time. 'About two metres,' was the reply, 'it will be OK.'

We returned to the hotel. It was called the Fisherman's Hut (the 20-letter word telling me this was unreadable to me in its Finnish form), but believe me this was no fisherman's hut as we might call it: big, luxurious and expensive, very warm and it appeared to be the centre of the locality's nightlife. Daylight in February seemed to arrive about 10.30 am and went away at 2.30 pm so the nightlife could be described as long. A wonderful dining room, with reindeer, elk and goodness knows how many kinds of fish on the menu and a cabaret . . . I decided Schindler were doing me proud even if I was not getting paid.

Nicholas and Michèle, married by this time and living in their own house, also were directors of our company and were contributing a great deal to our success, Nicholas still delivering units at a high rate of knots. Sometimes we even needed to bring in a spare-time driver to deliver with another vehicle as well, but were well aware we might be living on borrowed time as far as our market went. However Ford had shown no indication as yet that they were about to enter the four-wheel-drive market. Now things really began to hot up.

Our market was improving and as I drove around the dealers it was amazing how many Schindler-equipped Ford tractors were about, I was told that Mr Wild would be coming over and was asked to meet him at the airport and take him to see Ford tractors and later County Commercial cars. This had me worried but I was assured that while our position as sole importer was now unassailable, there was a possibility that Ford would need to use the Schindler kit on one tractor model. So here I was back at Basildon again and in the executive suite for lunch discussing the importation of 200 units into the UK which were to be imported by Ford and fitted by County to Ford six-cylinder tractors. Soon we were in the County design office talking about fitting the kits in County's factory. Since I was their biggest competitor at this time, it seemed a little odd to say the least; but they were very pleasant people and even today I would never criticise County (the old company, that is). I am just sorry I was never working with them enough. I was present when the first kit was fitted and helped Ford to work out a parts and warranty system on the kit so I felt that Schindler's help to us in the previous years was being nicely repaid. It seems to me that the end of Roadless and County as tractor manufacturers, as I had known them, began at about this time.

About the time we started Farm Tractor Drives and produced our first four-wheel-drive kit using the Italian Sige axle, I tried hard to suggest that a kit to fit a Ford 5000 tractor should be strongly marketed by Roadless; in fact I obtained a provisional order for 300 from Ford France.

Originally the Sige axle had been one of the basic products sold and I believed there was still a large market for them; and now we were proving just how big that market was. However, Roadless and County had decided the future lay in producing complete four-wheel-drive tractors. I believe this was the main cause of both companies' downfall once Ford introduced their own four-wheel-drive tractors made 'in house'. Of course I was not privy to all the problems they had but do feel they might have had less trouble by recognising the limitations of the Ford conversion market sooner.

Next panic was a visit to Sweden. (Another English-speaking country?) Again a plane to Amsterdam and another to Stockholm. This trip was notable because it meant a visit to the Ford factory in Stockholm and, yes, I did meet two people who had been on the Tracteuropa caravan with me all those years before. Also a plane flight to a town in central Sweden, Östersund, a winter wonderland indeed – but so cold! On my visit of over 20 years before we had clubbed together and hired a plane to fly us into the country of the Laplanders – a seaplane, that is – and I well remember skimming across the lake to get lift-off and looking at the logs floating on the water kept off our path by just a simple chain and floats . . . . But not this time; it was covered in rough ice so no flights off the lake now. But is was a place of wonder. At night, walking from my hotel through the clear frosty air and snow-banked footpaths on the way to a good French restaurant, I could see, on the hillside opposite, people skiing, long sweeping runs and floodlights, looking so beautiful in the night.

Soon after this trip Mr Wild was again in England. We visited Massey Ferguson, who required 4WD kits to suit the model 265 tractor for export, mainly to Norway. We also visited their Trafford Park factory regarding the Schindler axle fitted to the 50D backhoe unit. Later we called on two other companies in the industrial field, so in the space of a year I went from being sole low-volume importer to becoming their representative with main manufacturers who, in total, were ordering 1,800 units. It was not easy. I was still competing in the low-volume kit market with most of the manufacturers I was visiting for Schindler.

I mentioned before that this market was full of politics, mainly due, I suppose, to the lack of people who had the knowledge to cover the many fields required without training; thus many of us kept meeting up working for yet another company. One of our industrial customers called me one day to say we must get Mr Wild over very quickly – they had a steering arm break and the tractor had run out of control and partly demolished a parked car and what were Schindler going to do about it? Did we realise the seriousness of the situation if someone had been killed or injured and how there was a risk of the Health and Safety Executive throwing the book at us? Mr Wild agreed to come and later that week we presented ourselves at the factory. Grave faces sat around the table, two or three people explained to us how seriously they were taking this matter and the question 'What are Schindler going to do about this?' was repeated. 'Nothing,' Mr Wild replied. As the temperature rose he was threatened with lawsuits, loss of trade, etc. 'But how can you do that?' he asked. 'Your buyer felt our unit was very expensive and that steering arm, being a simple fabrication, was deleted from our specification because he said that it, and several other simple parts, could be made at lower cost by you. So what are *you* going to do to protect *our* good name, because we may be tarnished by association?' We went away and treated ourselves to a good lunch. We were not offered lunch by that company, but they did buy many more axles without steering arms.

I knew Schindler had several contacts in America and our efforts to sell over there were not very successful. I had not enough time to devote to the project. One of the Schindler contacts decided to send two people over to the Paris Show to look at 4WD options, and they were to visit Switzerland before the show. I was invited to be present. A good meeting followed and it was arranged that I would visit America to supervise the installation of the first kit, as soon as the ten kits they had ordered arrived.

During this visit to Switzerland I found it was what the people of Basle call 'Carnival' and so we were taken into the city to help celebrate. It was a wonderful event: the whole city seemed to be walking the streets, frosty air, big bands, small bands, even groups of mother, father and children walking around the town playing flutes and drums; really it seemed as though everyone in town could play a musical instrument. The air of excitement and sound of all this music just had to be experienced to be believed. The air was alive. I asked 'Why have a carnival now, in the middle of winter?' and was told that the people of Basle were such good Christian people that they had special dispensation to use the first week of Lent for a carnival before they started Lent proper – and this certainly was the first week of Lent. Maybe they were kidding me, but it certainly is one hell of a carnival, starting Monday morning at 3 am and continuing with music, floats and celebrations until Wednesday night. All night and all day; we left at 4 am and there was no sign of it ending. We returned on Tuesday afternoon and it was still going strong; the many bars were still open and would not close until Wednesday evening. We walked around, calling in more bars than I care to remember.

We had bought a special badge for about £5 that guaranteed us immunity from the characters that wander town collecting money for the many charities of Basle, and if you do not have a badge or do not give a donation to them you are likely to get a handful of confetti stuffed down your neck. The streets were awash with confetti when we retired to our hotel in the early hours of the morning, but next day at breakfast time the whole town was clean again and festivities still going on as we left. What impressed me most about this carnival was the total lack of police presence coupled with a complete sense of security at any time, no yobs, no embarrassing drunks, just complete good spirits. I have been back several times and found it always the same.

When the kits were ready after having special gears made to suit the American Ford tractors, which tend to have different tyre sizes from those we usually see in Europe, I left for America, arriving in Chicago about 4 pm. I had a hire car arranged and left the airport for the 250 or so mile journey to my destination. After driving about two hours on the interstate highway that runs around Chicago and passing continuous housing estates and built-up areas, I found that I was still by-passing this great city. The radio was still warning me about traffic problems at the junction of 12th with Dayton Avenue, or the traffic lights were not working at another junction . . . true, I was driving at 55 mph as the speed limit decreed. I knew America was big but never till now did I realise just how big. After two and a half hours I reached the junction of interstate roads and turned off into Indiana, heading for Lafayette and my Howard Johnson Hotel.

The next morning at Mud Hog factory, a company specialising in making 4WD conversions for combine harvesters, there was a board outside, announcing 'Welcome to our guest of the day, Mr Battelle from England'. I really had arrived in America! Already the receptionist in the hotel had told me 'Have a nice day!' Wonderful country . . . and how I love going back! It was good to meet all the management at Mud Hog, the only snag being that our equipment had not yet arrived, but it worried me not, I was having a great time. That evening I was introduced to American steak, prime rib. Did I want a 16 or 32 oz steak? Settling for 16 oz, it arrived with lobster pieces as dressing. A wonderful meal in the Clock restaurant in Lafayette. (The clock I found had been imported from a Lancashire cotton mill that had met its doom some years before – a much-travelled timepiece.)

Staying in this hotel on another visit the more rugged side of Indiana was in evidence; a telephone call to my room asked that I go down to the ground floor as a tornado was approaching, and looking out of the window I could see this black spinning column approaching. With roaring ferocity it must have passed close by and then the rain came, a solid sheet of water. After a short time people began arriving in cars and taxis absolutely drenched, water ran in puddles onto the floor as they walked. The other Howard Johnson Hotel across town had been partly demolished by the tornado. Indeed there can be two sides to every country.

I was amused on one visit when I discovered one of the receptionists came from the South, with such a lovely southern drawl. When I asked for a room she looked at me and asked where I was from. 'England.' 'Gee, Ahh could listen to yo talk ahll night', she said. She did not have the chance but did come to visit us in England. I am always sorry we were on holiday and she never found us.

Returning home brought me back to earth with a bump and the urgent need to find something to market to replace the Schindler connection. Although we did not realise it, our Schindler market had only about eighteen months to go. Just as Massey Ferguson were about to order another batch of axles there came a problem in Norway, so again I was off to an 'English-speaking' country. The problem was in the hardening of the gears and so it was decided to change the driveboxes and Schindler agreed to make a batch of 30 available so the Massey importer could change them, sending the old ones back to us to be rebuilt because there was a good transport system between UK and Norway. What surprised me was to find the tremendous distances involved in moving around Norway; the importer told me it would be nearer to ship a box to Milan in Italy than to his northern dealers.

The shipment to Banner Lane in Coventry to meet the Massey Ferguson order started to arrive and I was constantly called to answer the inevitable queries that always go along with new business and new customers. Massey were constantly reviewing their requirement for kits and increasing the estimated volume, as indeed we were, and Ford wanted extra supplies. The dealers in Europe also were making more demands on the Schindler factory. On the surface it seemed to be a wonderful opportunity to do business but we knew it to be a fool's paradise; at this rate the requirement for 4WD tractors would soon mean the main manufacturers needing to have this facility in their own production and so our estimate of how long we had to operate this lucrative market before the end came was constantly being shortened.

There were several more trips to Norway for me. We had a backhoe manufacturer interested in using our kit to offer his customers a 4WD backhoe and on one occasion I visited him during the winter and had my first experience of driving on studded tyres. I had hired a Ford Fiesta to drive about 40 miles north of Oslo. The snow was deep and hard packed, no grit or salt being used, and I was driving very carefully. It was not yet daylight when I saw a lorry catching up quite quickly, it soon swept past, a long articulated unit with chains swinging along the sides of the body, obviously a timber truck on its way to collect a load . . . a roar, cloud of powdered snow and it had gone. How the hell does he keep that on the road? I thought. At the next bend, never slowing, it went round like a truck on rails. I expected it to slide or jackknife or for some other disaster befall it, but no, it just left me standing. Must be the tyres, I decided. The road surface was such that if it had been in England there would have been cars in the ditch, up lampposts, sideways in the road and all the traffic soon at a halt. I applied the brakes carefully; the car slowed without a trace of a slide. Building up more speed, 50 kilometres an hour, I applied the brakes again, quite hard; the car's nose went down under the braking effort but not a trace of a skid. My first experience of using snow tyres. When I stopped at the next town it was difficult to stand on the road and, feeling the tyres, there were lovely little spikes sticking up over the whole surface.

One day a message came from Ford in America to see if Schindler were interested in quoting to supply 4WD kits to them for a new project and if so would they come to Detroit for a briefing on the requirement. Schindler were interested and so Mr Wild and I took off for Detroit at the appointed time. We were told of the specification Ford required and the delivery date (which looked to be impossible) and sent on our way. We visited Mud Hog to check on their progress and they told us of inquiries they had. We visited both Case tractors and Allis-Chalmers, after which Mr Wild returned to Switzerland and I returned to Mud Hog to assist. This culminated in some visits to customers and finally trying to work out a new gear ratio system for a specific need on a pea harvester.

It looked as though I would miss my flight back from Chicago, but Mud Hog offered to fly me to Chicago if I did not mind flying in their aircraft which was only single engine. I knew they used it often and had a professional pilot so I reckoned it would be OK – in fact something new for me to experience. They really excelled themselves and obtained clearance to land at O'Hare Field instead of the more normal private airfield nearby, so we were actually landing at the international airport where my return flight to UK would take off from. It was a great experience flying over the American Mid West with its square fields, lovely white farmsteads and eventually the miles of built-up areas as we approached Chicago. Dave the pilot had put me in the co-pilot's seat so I could watch him operate the plane – very interesting indeed. 'Do you want to listen to airport control?' he asked while talking us in. 'Yes, please.' Anything new must be interesting, I thought. O'Hare is one of the world's busiest airports and we were told to circle at a specified height. After about ten minutes the instruction came to approach the strip. Ground control told us: 'You are following a jumbo in so keep well back or the turbulence may cause you danger.' Dave did as he was told, we followed the jumbo and could see another twin-engined passenger aircraft in front of it. Ground control was talking to this aircraft telling him to hold back a little, as the runway was not quite clear. We could hear this very distinctly. The controller's voice took a sudden slightly harder tone, obviously talking to the plane in front of the twin-engined jet: 'Get off that goddamn runway or you will have a 727 on your back.' Dave laughed. 'Never changes,' he said. We followed the jumbo in at a respectful distance landing safely and smoothly, but it seemed quite incongruous that three jumbos were waiting to use the runway whilst our tiny plane taxied in front of them.

Back home our accountant was having a battle with the tax man who was finding it suspicious that we had grown so fast without borrowed money. Probably he was thrown by the fact we could pay his demand by return of post; and so long as our bean counter agreed with the tax man's demands we always did pay quickly. This seemed to throw him but I still felt that if there were no debts to pay, what money we had was ours, so we did not need to make obscure estimates about how we would meet bills in the future. We just had the money now and if we needed to spend, then without question we could.

The Paris Show came round again. It was to be our last, but I did not know this at the time. We met as usual and set out to do business and enjoy the nightlife. In Paris one can find all kinds of entertainment. One evening we visited a small restaurant near the Sacré Coeur Cathedral . . . nice meal, good live guitar music . . . we could talk, sing and admire the restaurant ceiling completely covered in banknotes, probably old ones, but interspaced with the occasional pair of ladies' panties (I guess those represented particularly good nights in the place). Another night we visited a restaurant we had been told about on the Île de la Cité near Notre Dame – why do we keep going near to cathedrals? Anyway we found the restaurant with some difficulty, down a passage and into a cellar. Yes, they had a table, gave us a number, directed us to it and gave us a glass each telling us to help ourselves to wine as we passed the barrels at the end of the passage, our waitress would be with us shortly. Two barrels of wine were available, red and white, no choice just get on with it. Our waitress arrived, offered us chicken or steak; bread was on the table and the starter course was salad. Had we done right in coming here? Brick cellar with various arches and tables, one central arch housing a trio playing popular music, I guess probably students earning a bob. Our salad arrived, in a galvanised bucket, ice in the bottom, celery, lettuce, peppers, some things that tasted OK but I knew not what they were, tomatoes in a dish, plenty of bread, help yourself to the wine, the meat course, no veg, cheese board afterwards, help yourself to the wine, would you like coffee, help yourself to the wine. . . . We had a good evening.

The next night we reserved a table at the Crazy Horse Saloon, a floor show giving us an hour's entertainment, good value for money. Afterwards we visited a night-club . . . another floor show, a few bottles of champagne at £25 a bottle and we gave up the nightlife. . . . Why were we here? Oh yes, a tractor show. . . . Good memories, but it was not all good living and money. Next day looking round the show the seeds of an idea came to me that might just find Farm Tractor Drives something to sell.

Back home talking it over with the family, a board meeting if you like, we identified a specialised market for a tractor. The current list price for 100 horsepower tractors was around £19,000, but if we took a used Ford tractor available at £2000 to £3000, without Q cab, sold the old engine and front axle, rebuilt the hydraulics and rear transmission (they never needed much money spent on them), fitted new six-cylinder engine, Q cab, Schindler kit, new tyres, rewired the unit, new battery and our own tinwork to establish an identity in the market, it appeared we could do this for £7000, allowing the dealer to take £3000 and us to take £3000. In short, an almost new tractor for £13,000. I soon drew up dimensions for side members to support the new six-cylinder engine and front axle and commissioned six sets to be produced. The Duncan Cab was legal to use if it would pass a noise test. It did, so we had a Q cab with radio and heater ready to go.

We soon had the first unit built, the final retail cost to be £13,500. This was our demonstration unit. The parts we did not use from the old tractor were sold for £600; and, best of all, because we had new side members and front axle support and were using our own serial number, which made a new registration number obligatory, we thus got over what had been my only reservation about the project – would customers pay this amount of money and still end up with an old registration number? Now they could produce a new tractor for their neighbours to see with a new registration number. In fact basically it *was* a new tractor because we were giving a year's warranty on the complete tractor including the rear transmission which we had rebuilt. A leaflet was produced and press release sent to all the people I knew who might give us publicity and the results were encouraging – we were featured in several periodicals and put the tractor on our summer showstands.

In the meantime we had made another trip to America to present a cost-and-design specification to Ford. This time Nick accompanied me and after seeing Mr Wild off on his return journey to Switzerland we took a flight to San Francisco. In the nearby Napa Valley there was a dealer who had taken delivery of some 4WD units from Selene before Schindler had bought that company but who had never paid for them. It was time to find out why. We did – he had gone bankrupt two years before. So registering a claim with the receiver we had to leave the matter, but we enjoyed the Napa Valley, visiting a book publisher whom we had bought books from in the past and who had a vineyard; also we enjoyed the city, visiting its famous places, eating on Fisherman's Wharf and calling on Jack Heald, the man who ran the Fordson Tractor Club in America. He took us to see several Fordsons and have interesting chats with their owners and next day we visited a most interesting museum owned by Mr Heidrech with its acres of old tractors and machinery awaiting restoration and also to see the many machines already restored. He had several old Holt crawlers steered by a single front wheel, massive machines and I guess the forerunner of the famous caterpillar tractors. He also had a Holt combine built in the early part of this century and made almost entirely of wood, reputed to need 24 horses to pull it and having a walkway across the rear of the cutter platform to help in case of blockage and to raise and lower the reel (sails), which was accomplished by operating a long spar with a counterbalance on the end of it. He also had a Fordson model F not running on wheels but which had a kind of screw thread on each side of it so that when turned by the engine it screwed itself along – apparently very good in snow.

We flew back home to a never-decreasing demand for 4WD kits but now felt certain this would diminish within the next year. I still felt there would be a demand, although on a decreasing scale, because there were enough two-wheel-drive tractors being sold to provide a good market for conversions in the immediate future but this would be limited. *Fordson Magazine* had been expanded to include other makes and our vintage tractor spare-parts market was growing. Also the Schindler parts business was becoming better now that some of the first kits we had sold were reaching as much as 10,000 running hours.

During one of the visits to Switzerland Mr Wild told me Ford would not purchase Schindler equipment for the new models unless the cost could be significantly reduced, also that Massey Ferguson would produce their own unit in a year or so, after the latest manufacturing schedule had finished. The Swiss franc was increasing in value so it seemed most unlikely that Schindler would, or even could, reduce the cost of the units to meet this competition and thus Mr Wild felt it likely that manufacture would cease. This represented a blow to us but not an unexpected one.

Another trip to Norway materialised. It was to be the last one as it turned out, but two things stick in my mind about this one. Firstly I was booked into a hotel the name of which I shall never forget: it was Hotel Hellzapopin – not pronounced like that by the Norwegians but a never-to-be-forgotten hotel, very swish indeed. The other experience drew my attention to the laid-back lifestyle in Norway at that time. During my wait for the flight to be called in Oslo Airport a very well-dressed man casually walked through the departure lounge, preceded by a single policeman wearing a revolver, and he took a coffee from the refreshment counter. I had been chatting to the man sitting next to me and asked 'Who is that?' 'Oh, that is King Haakon,' he said. No one took any notice of him – so different to our situation.

Back home our tax office were as usual, being a pain, wanting information and documents. Our accountant, now a friend, was doing his best but still obviously charging for his services, so we now had to spend time finding documents and thinking back, perhaps two years, to answer the tax office questions – and at our pace of life, two years ago was almost prehistory to us. If we had have been trying to defraud them I should have enjoyed pitting my wits against them but we were so busy with so little staff that there was no time for those kinds of luxuries. Our office system was so simple: we recorded every transaction and simply filed the paperwork away in monthly files. Not the best system but not the time-wasting one that seemed to be understood by the tax office; but that would have needed more staff and the tax office scrounged our time enough as it was, without paying more staff for their benefit. We were spending too much time on unpaid work for the Government, collecting income tax, insurance tax and VAT for them. A rough estimate told me this service was costing about £3000 each year. They must think all business people are daft. Well, perhaps they are, many still do it. A very nice lady arrived from the VAT office to check our accounts. We had had a previous visit some years before and nothing was found wrong so this lady was only concerned with the last three years. At the end she was very apologetic and told me she had some bad news for me. Apparently we could not claim back VAT we had paid in transporting our kits from Switzerland, as only the bit paid for transporting them in the UK could be claimed. The previous inspector had not picked this up but this lady had done so. Even our accountant did not know this. I reckoned the Government made the rules up as it went along; however we did not blame the lady. She told me the amount was £2000 and gave me the idea that she had doubts that a small company would be able to pay this amount. I never made any problem of this and just accepted what she told me and agreed to pay. We soon received a demand from the VAT office and sent them a cheque. That was the final straw. I decided no more money would be paid to a Government hellbent on supporting scroungers and giving money away to far-distant countries who seemed to do nothing but buy arms with it to fight civil wars. I would put myself in a situation where the tax I paid would, quite deliberately, be the least possible whilst staying within the law.

We were fortunate that about this time a Ford tractor dealer showed some interest in purchasing our company and I started to talk very seriously to him. In the meantime we received an income tax bill for £20,000 – a legitimate bill. There was no query on my part other than the thought 'Who the hell am I working for?' Our accountant was horrified at the sum, but nevertheless we had earned the money so I could have no legal complaint. However, he suggested we start a pension fund. By sheer chance at that time the maximum amount it was possible to pay into a fund per year was £10,000 so we made one payment on the last day of our financial year and the next on the first day of our following financial year. This, added to the small existing one we had, has stood us in good stead; the tax people understood this and never made a murmur about it but I am sure they were always unhappy about our method of business – it was too simple for them to understand. I laugh out loud now when I hear the politicians bleating about helping small businesses. . . . Am I bitter? I suppose it sounds like it but I *feel* no bitterness, I have had a good life and paid my taxes more or less willingly and at least have slept well at night, give and take the odd nightclub.

Our tractor, which we called the FTD Chieftain, was now getting good publicity and inquiries were coming in; we had completed three units and other companies were showing interest in taking parts from us to build their own tractors. This was something I had not expected but could have been a good money-spinner. By this time the tractor market was becoming more competitive. We had expected this and believed we could still hold our own in the market, but what did take us by surprise was the deep price cuts by the big manufacturers, who were now in a deeper sales war than I could ever recall. It made me wonder how much profit they had been making in the first place. The normal discounted price for a 100 horsepower tractor was down from £19,000 to about £15,000, which suggested the price advantage we had originally envisaged of about £5000 was nothing like that now. Indeed, initial estimates gave the idea that our price advantage had shrunk to £1500 or so, and I was very careful not to become overstretched by putting too many parts into stock.

Talking to a few dealers who had been interested in our project soon gave the impression that the FTD Chieftain must die or we would be working for no reward. It seemed a great pity because the tractor had proved a good worker and because it seemed to have the weight distribution just right, which is what I intended in the first place. But if the sums did not add up there was no advantage in carrying on. However we could probably have sold quite a good number of kits to small companies and farmers, as well as still having a limited market in four-wheel-drive kits to be fitted to tractors already on the market. Before this situation could be properly assessed the decision was made easy for us.

The Ford dealer who had shown interest in Farm Tractor Drives came up with a firm offer and we accepted, so our connection with Schindler was at an end. The new owner needed Nicholas to help him and I have often joked about selling my son with the business.

It was now June 1984 and all I had left to occupy my time was the vintage tractor spare parts business and *Fordson Tractor Magazine*. Our staff were told they would be paid until September 1984 but could leave any time if an alternative job turned up. It was very pleasant to enjoy an easy summer. Nicholas began to sell Ferguson 20 spare parts from his garage; as I remember, he started with only £100 of spares in stock so over the years The Old Twenty Parts Company has grown considerably. Eventually he tired of working for someone else and left the Ford dealer to devote himself full time to the spares business and so he took over the Vintage spares business I was running, along with the premises. Our Iron Warehouse was almost at the end of its lease so that was not renewed. *Fordson Tractor*

Magazine was sold and Betty and I retired in 1985, quite early and in line with my determination to stop giving the Government too much tax money and (worse still) collecting it for them for free.

Looking back over the years covered in the three books which I now call my Tractor Trilogy, I note some things are not yet recorded. My father is no longer mentioned towards the end of the third book. Unfortunately by the time this story has reached its end he was no longer with us, having passed away relatively peacefully after a four-day illness. Looking back over his life I am struck by the similarity of his life and mine. My father started work on a farm and was never able to amass enough money to go into business until he was 40 years old and then with considerable help from my mother, when they started the haulage business described in the first book. This ran until they had the opportunity to rent a farm, a long-standing ambition which gave them both satisfaction and the joy of achievement.

I was 35 years of age before Betty and I became even remotely prosperous, and over a period of 25 years, about the same time span as my father had a farm, we achieved prosperity allowing us to retire. Our son Nicholas and his wife Michèle started in business many years younger than Betty and I did and have already exceeded our achievements in many ways. Only the future can tell how far their achievements and ambitions will take them. And what of our grandchildren Clare and Philip? Only time will tell.

Perhaps you will have another book to read written by another author in the years to come, but if they can tell you of a life as interesting and good as Betty and I have had, they will be lucky indeed.

Farming Press Books & Videos

Below is a sample of the wide range of agricultural and veterinary books and videos we publish. For more information or for a free illustrated catalogue of all our publications please contact:

Farming Press
Miller Freeman Professional Ltd
Wharfedale Road, Ipswich IP1 4LG, United Kingdom
Telephone (01473) 241122 Fax (01473) 240501

Books

Early Years on the Tractor Seat **Arthur Battelle**
More Years on the Tractor Seat
The first two volumes of Arthur Battelle's informative and humorous 'Tractor Trilogy'. *Early Years* covers his youthful passion for agricultural machinery and experiences on the land during the Second World War. *More Years* takes him from cultivation contractor and agricultural engineer to tractor demonstrator at home and abroad.

Tractors at Work **Stuart Gibbard**
Vols I & II
Each book contains some 180 photographs spanning 1904 to the present and showing a wide range of tractors in many working situations on farms in Britain.

Ford Tractor Conversations: the story of Doe, Chaseside, Northrop, Muir-Hill, Matbro and Bray **Stuart Gibbard**
Detailed, profusely illustrated account of the main models and machines produced by these leading companies.

Videos

Classic Farm Machinery **Brian Bell**
Vol I 1940–70
Vol II 1970–95
Archive film extracts tracing the mechanisation of the chief arable operations.

Classic Tractors **Brian Bell**
Archive film extracts focusing on the development of tractors from 1945 to the present.

Fordson, the Story of a Tractor **Michael Williams**
The Massey-Ferguson Tractor Story
John Deere Two-Cylinder Tractors (Vols 1 & 2)
Henry Ford's Tractors 1907–56
Videos showing in detail the machines produced by these leading companies.

Farming Press Books & Videos is a division of Miller Freeman Professional Ltd which provides a wide range of media services in agriculture and allied businesses. Among the magazines published by the group are Arable Farming, Dairy Farmer, Farming News, Pig Farming *and* What's New in Farming. *For a specimen copy of any of these please contact the address above.*